The Manufacturing Arts in Ancient Times

MANUFACTURING ARTS

IN

ANCIENT TIMES

WITH SPECIAL REFERENCE TO BIBLE HISTORY.

BY

JAMES NAPIER, F.R.S.E., F.C.S., ETC. ETC.

AUTHOR OF
"MANUAL OF THE ART OF DYEING," "MANUAL OF DYEING RECEIPTS,"
"MANUAL OF ELECTRO-METALLURGY," "ANCIENT WORKERS IN METAL,"
"NOTES AND REMINISCENCES OF PARTICK," ETC. ETC.

"There is no new thing under the sun, the newest thing having been
already of old time which was before us, and being new only because there
is no remembrance of former things."—*Bible.*

LONDON: HAMILTON, ADAMS, & CO.
EDINBURGH: MENZIES & CO.
GLASGOW: HUGH HOPKINS.
1874.

16

PRINTED BY BALLANTYNE AND COMPANY
EDINBURGH AND LONDON

PREFACE.

—◇—

HAVING, many years ago, observed the frequent
mention in the Bible, and other ancient writings,
of articles which, for their production, must have
required a very advanced knowledge in the manufacturing arts, and also that incidental allusions
were sometimes made to the processes by which
these were produced, I became interested in the
subject, and started an inquiry as to how far such
passages should lead us to conclude that the
ancients were acquainted with processes similar
to those which we at present practise.   I commenced by making a collection of all the passages
which I met with in my reading having any bearing on the inquiry, especially those relating to
metallurgy. Having a practical knowledge of this
branch, acquired during several years' experience,

I soon came to form some definite conclusions on the subject, and as I considered these of some little importance, and likely to be generally interesting, I embodied them in a paper which I read before the Glasgow Philosophical Society, and subsequently I published these notes in an enlarged form in a small book entitled, "Ancient Workers in Metal."

As this book has been out of print for some time, and as since its publication I have considerably extended my inquiries, so as to embrace other trades and manufactures, of some of which, such as dyeing and weaving, I have also a practical knowledge, I now put the results of my inquiries before the reader in this volume.

The method of argument followed in this volume is the same as that which I pursued in "Ancient Workers in Metal." I argue that when we find mention made in history of certain manufactured articles as in common use in ancient times, we are compelled to concede that manufactories for the production of such articles must also have existed; and knowing the processes we now follow in pro-

ducing similar articles, we may conclude that the
ancients followed similar processes, at all events
processes embracing the same principles. And
wherever I have found the data sufficient, I have
endeavoured to show this similarity, in some cases
I might say identity, between the ancient and
modern processes.

This field of inquiry, I think particularly suitable for practical men, who, like myself, may
not have had the advantage of a classical educa-
tion, but who know the processes and practical
difficulties of their own trade. They are well cir-
cumstanced for reasoning upon such allusions as
occur in Scripture and other ancient writings, to
matters bearing upon their own branch of manufacture. Such a course of inquiry would open up
to working men a new source of interest in the
study of the Bible, and probably the conclusions
arrived at might eventually prove a service to
those whose profession it is to explain the meaning and truth of Scripture.

JAMES NAPIER.

PARTICK, *April* 1874.

CONTENTS.

———◆———

MANUFACTURING ARTS IN
ANCIENT TIMES.

———◇———

INTRODUCTION.

THE object which we have in view in the following pages is to ascertain, from such references as occur in the writings of ancient authors, especially such as are to be found in the Old Testament Scriptures, the probable state of the practical arts in very ancient times.

We intend viewing our subject from the standpoint of a practical workman, and we hope to show that long ere Greece had attained the rank of a nation, and thousands of years before she reached her artistic greatness, many ancient nations had attained to great, and even, when compared with our present excellence, astonishing proficiency in many of the industrial arts, especially in metallurgy, which may be considered the basis of all the industrial arts; also we hope to show that such knowledge and skill, though forced by

▲

circumstances from one locality, found lodgment in another, and so was never entirely lost to the world. By overlooking this fact, antiquarians have drawn many wrong conclusions; for instance, they have too frequently regarded the attainments of some particular nation at a certain era, as the index of the world's attainments at that time, forgetting the possibility that the world's advance may not have been uniform for the whole.

General history seldom gives details of manufacturing processes; nevertheless, the incidental mention of a manufactured article is evidence that such manufacture existed; and knowing what we would require to do to produce the same article, we may reasonably suppose that it would require some such similar process then; and if they did not possess the same facilities which we have, and if such articles, found by archæologists, be equal in material and workmanship to what we can produce, we are compelled to appreciate their skill as excellently great. For instance, when we find it stated in the Bible that certain articles were made of bronze, and know that this alloy is never found native, but is always the product of a manufacturing process, and when we know further that the manufacture of several of the articles named is attended with special difficulties, we are forced to conclude that the ancients had methods of overcoming these difficulties, and knew something of the laws of chemical action, and that therefore they had great knowledge and skill in respect to this particular art. By following out this method, and applying it to a variety of works of art named in Scripture, we will be able to appreciate better the meaning and value of the good old Book, and see that it is of the highest importance, as a history as well as a religious guide to mankind. This method of inquiry would make the Bible a very pleasant occupation, and one well suited for thoughtful working-men; and Biblical scholars would also gain, if practical men would thus throw their side-lights on the meaning of many passages relating to the arts, concerning which, at present, Biblical critics, as a rule, display a woful ignorance.

We have stated that the discovery and application of metals is really the basis of the mechanical arts, for it puts into the hands of man proper tools, whereby he is enabled to overcome difficulties which he could not possibly overcome without them; and indeed all advancement begins with improved tools; we will therefore, in the first place, consider the different metallic arts of which we have indications in ancient history, and then refer to some of the manufacturing arts that are of a chemical character, such as dyeing.

The origin of metallurgical operations can only be a matter of speculation, and the speculation will vary according to the position and imaginative faculty of the individual venturing the solution of such a problem. The wonderful results of the operations, and their importance to man, have caused many, both in ancient and modern times, to believe that the art must have been communicated to man by the gods. Others, again, have ascribed the origin of the manufacture of metals to accident; they have supposed that a certain mineral, by the force of fire, may have been made to yield a metal, and that by repeating the process with other minerals, other metals may have been found out, and ultimately all the different metals which lie concealed in mineral combinations in the earth.

Various ancient historians speak of silver and other metals being melted out of the earth during the burning of woods upon the Alps and Pyrenees; and even so late as 1762, a large mass of mixed metal, composed of copper, iron, tin, and silver, was melted out of the earth during the conflagration of a wood that was accidentally set on fire; such may have suggested the burning of certain stones, and the manufacture of metals in some localities; but this can only have been one of the ways by which metals were discovered. Several metals are found in a native state, requiring no operation to bring them into use, and this may have led to trials being made of different minerals, which were high in specific gravity. Whatever may have been the origin of the art of metallurgy, certain it is that it originated at a very early period of man's history; and the knowledge once obtained, man has carried with him to every part of the world, and with him down the stream of time till the present day. Like other arts, metallurgy has had at different periods, and in different nations, a special and temporary rise, prevalence, and decline. These recurring periods, together with the meagre records we possess of ancient history, make an investigation into the progression of the art from early ages a very difficult undertaking; but through the aid of researches now being prosecuted amongst the buried cities of the ancient world, we have hope, especially if careful analyses are made of the various metallic products found, that ultimately we may arrive at a very clear view of the condition of the metallic arts at different periods, and also a pretty correct idea of the progressive history of these arts.

In this inquiry particular attention will be given to the methods of extracting the metals from their ores; for although it be allowed that the ancients may not have known the chemical principles embraced in such operations, still the very fact that they did extract metals is proof sufficient of a very great amount of practical knowledge and skill on the part of those employed in and directing such work. In order that the reader may better understand the difficulties met with in the processes of extraction, and appreciate the skill required for overcoming such difficulties, it will be necessary to treat of each metal separately, describing the methods now in use for obtaining the metal from the ore, and the various means adopted for overcoming natural difficulties; and thereafter, comparing any references which may exist in Scripture or other ancient works, we shall see if such references are applicable to our present processes, or to methods known to have been in use, or if they are to be relegated to methods entirely lost.

The nature of the country producing the raw material of an art, determines the direction and extent of knowledge of the art which the inhabitants of that country will possess. For example, the inhabitants of a country which does not produce ores of metals will not, as a natural consequence, know much practically of the art of extracting metals from ore, except some other natural circumstance interfere to procure a reversion of this rule. We have an example of this in our own country. Cornwall and Devonshire are the principal localities for producing ores of copper. These countries, consequently, produce the most skilled miners; but, from the absence of fuel or coal—an essential substance in smelting—the inhabitants are not copper-smelters. Glamorganshire, again, which produces no copper ore, yet produces the most skilled smelters of copper ore, because of the extensive supply of fuel in that county, and its position to the ore-producing counties making it more convenient and cheaper to convey the ore to the fuel than the fuel to the ore. Similar circumstances exist in Cuba and other localities, so far as regards copper ore. But in all cases where both ore and fuel are found together, as a general rule, the locality producing them will also possess skilled miners and metallurgists, so far as regards smelting, or reducing the ore to metal; and the same circumstances will, to a certain extent, produce the same results in every age and every country. If metals are in demand, and the ores are found in one locality and fuel in another, means will be adopted either to carry the ore to the fuel, as is done in our country, or the fuel to the ore, as is done in Chili and Australia; and if both ore and fuel are found together, they will be smelted on the spot, and the inhabitants of that locality will be skilled in the several arts of extracting and mining. If, then, our object were the locality of metallurgical operations, our inquiry would be concerned with the physical character of ancient countries, which would lead to an inferential reply; but our inquiry being more concerned with the operations than the locality where such operations have been conducted, we will not give any prominence to the locality, except where it may illustrate the operations themselves.

We still know very little of the metallic products of Palestine. Travellers who have visited that country, and who have recorded their observations, have either been incapable of giving the information, or their minds have been too much occupied with other matters to allow them time to investigate the mineralogical character of the country. Let us hope that the Palestine Exploring Expedition, which has done really good service in acquainting us with the mineralogy of the desert of the Exodus, may at an early date return to work, and enlighten us concerning the mineralogy of Palestine itself. As an inducement to the children of Israel to leave Egypt, and take possession of Palestine, Moses, who doubtless had personal knowledge of the mineral-producing character of the country, described it as "a land whose stones are iron, and out of whose hills thou mayest dig brass" (copper). We have, therefore, reason for believing that Palestine was a mineral-producing country. The very fact that Moses held out the mineral aspect of the country as an inducement is worthy of notice; for it is an indication that metals were in those days highly prized by both Egyptians and Hebrews.

No one can read the detailed account of the structure of the ark, with its furniture, as given by Moses, but must see that there were amongst the Hebrews men well skilled in the art of fabricating metals, and by whom the idea of going to possess a country yielding metallic ores would be highly appreciated. Nevertheless, we have little evidence of the Hebrews cultivating the art of metallurgy to any great extent after settling in their own land. The skilled artificers who left Egypt appear to have died in the wilderness before obtaining an opportunity either for practising these arts, or teaching them to their children, who, along with their offspring, were for centuries afterwards so occupied in war—in taking possession of their own promised land, and providing for the more immediate wants of the community —that the arts, more allied to a people living in peace, were not only not practised, but the knowledge of how to work in them ceased. The practice, common then to victorious nations, of carrying off as prisoners all artificers, especially such as manufactured the implements of war, reduced the Hebrews during the wars of the Judges to a very low condition as regards the metallic arts.

Concerning the days of Samuel we find the following graphic statement : " Now there was no smith found throughout all the land of Israel ; for the Philistines said, Lest the Hebrews make them swords or spears ; but all the Israelites went down to the Philistines to sharpen every man his share, and his coulter, and his axe, and his mattock, yet they had a file for the mattocks, and for the coulters, and for the forks, and for the axes, and to sharpen the goads." When the kingdom was established in Solomon's time, and when he was about to build the temple, he was obliged to arrange with Hiram of Tyre for skilled artificers in metal to execute the work for that house, the workmen which David had taken captive from other nations and reserved for this great object not being sufficiently numerous or sufficiently skilful — a sad illustration of the degrading effects of war upon a people.

The metals mentioned in Scripture, and in the works of the most ancient profane writers, are only six in number; namely, gold, silver, copper, lead, tin, and iron. Mercury or quicksilver was known, and applied to metallurgical processes before the Christian era, but no direct mention is made of it in Scripture, nor any allusion which we consider applies to it, consequently it will not be considered. It has been argued, however, that as metallurgists in ancient times employed mysterious language in respect to metals generally, and as their language with respect to mercury was especially mysterious, for they regarded this as the mother of all the metals, and through the influence of which all metals were fructified, purified, and brought forth, that this metal may have been known to the Egyptians in the earliest ages, and also to the Hebrews, and that mercury may be referred to in Scripture though in such a mysterious way that we miss sight of the reference altogether. In support of this thesis, passages like the following are advanced : " Only the gold, and the silver, the brass, the iron, the tin, and the lead, everything that may abide the fire, ye shall make it go through the fire, and it shall be clean; nevertheless it shall be purified with the water of separation: and all that abideth not the fire ye shall make go through the water." This water, termed *"the water of separation,"* has been supposed to refer to mercury.

We quote the following from Sir John Petus' translation of the works of Lazarus Erckern, 1683 :—

"And we are assured that in Moses' time they had the knowledge of all metals, as may be read in Numbers xxxi. 21, where Moses taught the soldiers how the spoils of their *heathen enemies* were to be purified. Commanding (as from God) that all their gold, silver, brass, copper, iron, tin, and lead, and everything that endureth the fire (in the furnace, according to the Syriac), should be purified by fire, and then to be accounted clean; yet it is also said in that text, that it shall be separated by the water of separation—by which water, certainly, is meant quicksilver—because this doth *purify, cleanse,* and *devour metals ;* and so Dr Salmon calls it a *volatile juice* or *liquor ;* for nothing but *fire,* or that *quicksilver aquafortis,* can separate those metals. The water of purification of men was a distinct water from the water of *purification* and *separation* for metals, and the ingredients of the one are communicated to us, but the Holy Spirit thought fit to conceal the other from us."

Thus we see that this learned author acknowledges the want of information upon the nature of " *the water of separation,*" but he nevertheless is convinced that it was mercury, and that the Holy Spirit, in consistence with the general spirit of the age, kept it a secret. We think there is not the slightest foundation for this supposition. The water of purification was prepared by burning a red heifer entire, then collecting the ashes mixing them with water, and allowing this to stand. This ceremony was instituted several years before the date to which the passage quoted is referable, and all that is stated in that passage is, that the passing of the metals through the fire did not render them ceremonially clean till they were washed by this water of purification. Besides, mercury does not purify, neither was it used for separating copper, lead, iron, and tin, which metals are named with the silver and gold that were to be passed through the fire, and then washed in the water of separation.

The construction of the ark of Moses, generally termed the Ark of the Covenant, is the first work of art of which we have any details in history; it was also the first work performed by the Israelites as a nation. Few can read over the record of that undertaking, and take into account the adverse circumstances of the Israelites—wanderers in the wilderness—without perceiving that many of the Israelites possessed great skill in various departments of art; some of them had probably been the highest-class artisans of Egypt. Nor was this skill confined to the men; there were among the women many skilful weavers, embroiderers, spinners, and workers in other arts then common to females.

Moses, at a general meeting of the whole congregation, explained the plan of the ark and its construction, and proposed the means by which it should be carried into effect; this proposition was heartily responded to. "And they came, both men and women, as many as were willing-hearted, and brought bracelets, and ear-rings, and rings, and tablets, all jewels of gold: and every man that offered, offered an offering of gold unto the Lord" (Exod. xxxv. 22).

Moses then arranged for the construction and completion of the work, appointing Bezaleel and Aholiab as overseers and superintendents of the whole. "And Moses said unto the children of Israel, See, the Lord hath called by name Bezaleel the son of Uri, the son of Hur, of the tribe of Judah; and he hath filled him with the spirit of God, in wisdom, in understanding, and in knowledge, and in all manner of workmanship; and to devise curious works, to work in gold, and in silver, and in brass, and in the cutting of stones, to set them, and in carving of wood, to make any manner of

cunning work. And he hath put in his heart that
he may teach, both he, and Aholiab, the son of
Ahisamach, of the tribe of Dan. Them hath he
filled with wisdom of heart, to work all manner of
cunning work of the engraver." (Exod. xxxv.
30–35).

When reading this, and what follows of the
narrative, there are many who consider that these
men were instantaneously gifted, or inspired by
God, with a knowledge of their work, and that
they had not passed through an apprenticeship of
practical experience: such an idea we think
erroneous. Bezaleel was filled with wisdom; according to our interpretation, was a man of great
talent, having both head to understand and hands
trained to work. So was Aholiab; and it is evident from the record, that the one had a greater
knowledge of certain kinds of work than the
other. They were skilled because they had learned
these arts in Egypt. But all wisdom is from God,
and Moses speaks in the current language of the
people, and in accordance with the Hebrew philo-
sophy, which refers all results directly to God.

That these two were only overseers, or responsible men, and that they were assisted in the
performance of all this work by other Hebrews,
is evident from the following passage :—" Then
wrought Bezaleel and Aholiab, and every wise-
hearted man, in whom the Lord put wisdom and
understanding to know how to work all manner

of work for the service of the sanctuary, according
to all that the Lord had commanded. And Moses
called Bezaleel and Aholiab, and every wise-
hearted man, in whose heart the Lord had put
wisdom, even every one whose heart stirred him
up to come unto the work to do it. And all the
wise men, that wrought all the work of the
sanctuary, came every man from his work which
he made " (Exod. xxxvi. 1, 2, 4).

If we read carefully the account of the work
done, and consider its amount and variety, we
must form no mean conception of the art knowledge possessed by these Hebrews.

To prove that the work performed by these
Hebrew artisans in the wilderness shows no mean
knowledge of the metallurgical arts, we will par-
ticularise a few of the articles made by them. In
the first place, we notice the two cherubims and
mercy-seat. The particular shape and form of the
cherubim is unknown. It is generally agreed
that they had a human head, whether a body, and
whether of human form, or of some animal, is not
certain. The representations given of these figures,
which were common in Egypt at that time, have
a body. They are thus described in Exod. xxxvii.
6–10. However, they are all represented with
wings, and such appendages were upon those
placed on the mercy-seat.

"And he made the mercy-seat of pure gold :
two cubits and a half was the length thereof, and

one cubit and a half the breadth thereof. And he made two cherubims of gold, beaten out of one piece made he them, on the two ends of the mercy-seat; one cherub on the end on this side, and another cherub on the other end on that side: out of the mercy-seat made he the cherubims on the two ends thereof. And the cherubims spread out their wings on high, and covered with their wings over the mercy-seat, with their faces one to another; even to the mercy-seat-ward were the faces of the cherubims."

To make a figure, whether the body was that of a beast or man, or merely a head with two out-stretched wings, measuring from two to three feet from tip to tip, with the hammer, out of one solid piece of gold, was no ordinary work, and a work which few, if any, artisans of the present day could accomplish.

The next piece of work was the candelabrum or lamp-stand. Its manufacture is thus described (Exod. xxxvii. 17–22)—" And he made the candle-stick of pure gold: of beaten work made he the candlestick; his shaft, and his branch, his bowls, his knops, and his flowers, were of the same: and six branches going out of the sides thereof; three branches of the candlestick out of the one side thereof, and three branches of the candlestick out of the other side thereof; three bowls made after the fashion of almonds in one branch, a knop and a flower; so throughout the six branches going out of the candlestick. And in the candlestick were four bowls made like almonds, his knops, and his flowers : and a knop under two branches of the same, and a knop under two branches of the same, and a knop under two branches of the same, according to the six branches going out of it. Their knops and their branches were of the same; all of it was one beaten work of pure gold. And he made his seven lamps, and his snuff-dishes, and his snuffers, of pure gold. Of a talent of pure gold made he it, and all the vessels thereof."

" And this work of the candlestick was of beaten gold ; unto the shaft thereof, was beaten work : according unto the pattern which the Lord showed Moses, so he made the candlestick."

Such words speak for themselves ; but our practical readers will be apt to say, Why do such work with the hammer, when it would have been more easily cast—a process they were well acquainted with? The only answer we are prepared to give is, that it was done according to order. We have no doubt there were significant reasons for so distinctive an order which have not been referred to. Ages after this, work made by the hammer was more highly valued than castwork.

Concerning the value of the different metals used in the construction of the ark and its furniture, we may here quote the opinion of Dean Prideaux :—

B

"The value of the twenty-nine talents and 738 shekels of gold will be £198,347, 12s. 6d. The value of the silver contributed by 603,550 at half-a-shekel, or 1s. 6d. per man, will amount to £45,266, 5s.

"The value of the seventy talents and 2400 shekels of brass will be £573, 17s.

"The gold weighed 4245 pounds; the silver, 14,603 pounds; and the brass, 10,277 pounds troy weight. The total value of all the gold, silver, and brass, will consequently amount to £244,127, 14s. 6d.; and the total weight of these three metals will amount to 29,124 pounds troy, which, reduced to avoirdupois weight, is equal to fourteen tons 266 pounds."

If we take the Scripture narrative, and consider the talent as equal to 125 pounds troy, and the gold as worth £5475, and the silver as worth £342, 3s. 9d. per talent, then we find that the whole value and weight somewhat differs from the above.

"And all the gold that was occupied for the work in all the work of the holy place, even the gold of the offering, was twenty and nine talents, and seven hundred and thirty shekels, after the shekel of the sanctuary. And the silver of them that were numbered of the congregation was an hundred talents, and a thousand seven hundred and threescore and fifteen shekels, after the shekel of the sanctuary. . . . And the brass of the offering was seventy talents, and two thousand and four hundred shekels " (Exod. xxxviii. 24, 25, 29).

From the data above-mentioned we conclude the following, in round numbers :—

29 talents	730 shekels	Gold,.........	£160,107
100 ,,	1775 ,,	Silver,.......	34,422
70 ,,	2400 ,,	Bronze,......	
Reduced to avoirdupois pounds, @ 1s 4d per ℔.,..			485
			£195,014

The weights given will stand as under :—

Gold,....3655 pounds	Avoirdupois,...3007 pounds
Silver,12,574 ,,	,, 10,346 ,,
Bronze,..8850 ,,	,, 7282 ,,
Total, 25,079 pounds	Avoirdupois, 20,635 pounds

Equal in all to 9 tons 4 cwt. and 1 qr.,—the bronze weighing upwards of three tons.

It must be remembered that this includes all the metals used for outside work connected with the tabernacle, as pins for tents, hinges, sockets, curtain rings, &c. If we include the weight and bulk of the curtains, and all the necessary furniture, the number of Levites required to carry the whole from place to place must have formed an imposing multitude.

Whether the tabernacle was carried by the Levites, or was only, as some suppose, taken charge of by them, the real carrying being done

by asses or oxen, the sight must have been
imposing.

The ark with its contents was comparatively
light, for it was carried between spokes or staves
by a few priests; in Samuel's time a cart was used,
the whole being drawn by two milk cows.

That the Hebrews had such a quantity of metal
at this time strikes a casual reader with surprise;
but it must be remembered that, on leaving Egypt,
they had individually obtained from their Egyptian
neighbours jewels and ornaments of various sorts;
besides, we cannot suppose them entirely destitute
of such things, as they had but recently conquered
the Amalekites, &c., from whom they had taken
much spoil.   The manner in which the collection
for the ark was made, indicates that many Hebrews
had metals in great abundance, and the collection
made at the numbering of the people, mentioned
in the following passage, proves this : " Every one
that passeth among them that are numbered, from
twenty years old and upward, shall give an offering
unto the Lord.   The rich shall not give more, and
the poor shall not give less, than half a shekel when
they give an offering to the Lord " (Exod. xxx.
14, 15).   This refers to silver, and shows that its
possession was universal, and the demand such
that it was not above the means of the poorest,
and, at the same time, far below that of the wealthy.
It has been calculated, from the number who left
Egypt, that the collection would form a sum amount-

ing to about £38,000. After the formation of the ark, it does not appear that the Hebrews were called upon again publicly to exercise their skill in the arts. Their forty years' wandering rendered such a call quite unnecessary. During that time all those who had learned these arts in Egypt died, and with them, in all probability, passed away much of their knowledge and skill, and except in making weapons of war, and probably instruments for agriculture, the Hebrews, for several centuries, were so much engaged in taking and retaining possession of their land, that the higher arts of civilised life could not well be cultivated; consequently, notwithstanding the enormous wealth that had been accumulated in the time of David, when his son Solomon began to erect the temple (a work which their forefathers, when they left Egypt, could have accomplished without assistance), skilled workmen could not be obtained amongst the Hebrews —there were found none who could undertake the superintendence or do the skilled work of casting and working in metals. Although the higher classes of Israel were very wealthy, yet amongst them were but few artisans, and no Bezaleels or Aholiabs, no cunning workers in all kinds of gold and silver, blue and purple, &c.; such men had to be obtained from another country. There were, however, masons and carpenters, and such workmen as were essential in a country, even in times of war; but the higher class of artisans had declined

amongst them. David says: "Moreover, there
are workmen with thee in abundance, hewers and
workers of stone and timber, and all manner of
cunning men for every manner of work" (1 Chron.
xxii. 15). Many of these referred to were no doubt
Hebrews, but we think it more than probable that
a great many of them were captives taken by David,
and kept by him for this great work, which is
somewhat established by the following passage.
When Solomon is negotiating with the King of
Tyre for men and materials, he says: "Send me
now therefore a man cunning to work in gold, and
in silver, and in brass, and in iron, and in purple,
and crimson, and blue, and that can skill to grave
with the cunning men that are with me in Judah
and Jerusalem, whom David my father did provide" (2 Chron. ii. 7).

As we have said, and as here exemplified by
David, it was the policy and practice of ancient
monarchies to enrich their capitals with the most
costly of everything—with great buildings and
utensils and decorations of the finest and richest
quality.

These structures were almost wholly the product
of manual labour. In their wars the victors took
a large number of prisoners; the able-bodied but
unskilled prisoners they set to work at ordinary
manual labour; but the skilled workmen were
employed in their particular crafts, to enrich with
their work the dwellings of the monarch and his

nobles, and the public buildings of the capital. To the modern, the costliness and profusion of decoration in some of these ancient buildings is almost incredible, and when we come to accept the descriptions as facts, we are forced to conclude that in the manufacturing and decorative arts the ancients were far advanced. Mr G. Rawlinson, describing the ancient palace of Shushan or Persepolis, says, " The floors were paved with stones of various hues, blue, white, black, and red, arranged doubtless into patterns, and besides, were covered in places with carpeting. The spaces between the pillars were filled with magnificent hangings, white, green, and violet, which were fastened with cords of fine linen and purple to silver rings and pillars of marble, screening the guests from sight, while they did not too much exclude the balmy breeze. The walls of the apartment were covered with plates of gold. All the furniture was rich and costly. The golden throne of the monarch stood under an embroidered canopy or awning, supported by pillars of gold inlaid with precious stones. Couches resplendent with silver and gold filled the rooms. The private chamber of the monarch was adorned with a number of objects, not only rich and splendid, but valuable as productions of high art. Here, impending on the royal bed, was the golden vine, the work of Theodore of Samos, where the grapes were imitated by precious stones each of enormous value. Here, probably, was the golden plane-tree, a worthy companion of the vine, and here, finally, was a bowl of solid gold, the work also of the great Samonian metallurgist."

As was the palace, so would be the dwellings of the nobility. Compare this description with what is said in the book of Esther, of Shushan the palace, and the reader will find the truth of the one verified by the statements of the other. The same author in describing Ecbatana, an ancient city of Bible times, says, " The battlements which crown the walls were of different colours; those of the outer wall or circle were white, next black, third scarlet, fourth blue, fifth orange, sixth silver, seventh gold." This, says Mr Rawlinson, may seem mythical, but he adds, " The people who roofed their palaces with silver tiles, and coated all their internal woodwork with plates of silver or gold, may have been wealthy enough to make even such a display as Herodotus here describes, and there is every reason to believe that in Babylon, there was at least one temple ornamented exactly as the city of Ecbatana is said to have been."

From a consideration of the different metals known to the ancients, and of the relics of the metallurgical arts which have been found amongst the ruins of ancient cities, we have not the slightest hesitation in affirming that the ancients obtained their metals from the ores by fusion.   We also think it probable that practical metallurgy had its beginning in the first ages of man's existence in the world; at all events, in the earliest ages to which history refers, we find metals named as in common use. We have no details of the particular processes adopted for obtaining metals from their ores in these very early times, neither have we details of the practical operations in the manufacture of metallic articles. We may have no description of the forms of their furnaces, nor of the materials of which these were composed, but we have evidence that they had a great facility in constructing such. In corroboration of this, we may instance the short time taken by Aaron to cast the calf or bull. Were artificers in this age placed in similar circumstances, and called upon to make or cast such a figure, they would find it difficult to do the work within the time the Israelites did it; and except they had the necessary moulds and other instruments, this could not be done without the presence of the most skilled artificers. All these the Israelites would also require, and, if they had such things with them—and there is every probability they had—then our position is strengthened, that the Hebrews were, many of them, the skilled artificers of Egypt, and that in leaving that country, they had carried their tools with them.

Upon the subject of moulds we may quote the following from Wilson's "Archæology:" "Among

all the varied and primitive relics which have been
from time to time discovered, none exceed in
interest the stone and bronze moulds in which the
earliest tools and weapons of the native metallurgists were formed.   They have been found in
Scotland, England, Ireland, and in the Channel
Islands, exhibiting much diversity of form and
various degrees of ingenuity and fitness for the
purpose in view, some of which are of bronze and
highly finished.   In the Museum of the Society of
Antiquaries of Scotland there are casts of a pair
of large and very perfect bronze Celt moulds of
unusual size and peculiar form." Probably the
ancient Israelites carried with them, as a part of
the spoil taken from Egypt, a bronze mould of
apis.   Stone moulds were used in ancient times
before bronze moulds came to be used.   Dr Wilson
says on this subject :—

"But still more interesting are the ruder stone
moulds, in some of which we may trace the full
efforts of the Aborigines of the stone period to
adapt the materials with which they are familiar
to the novel arts of the metallurgist.   This is
particularly observable in a stone mould preserved
in the Belfast Museum.   It is polygonal in form,
and exhibits upon four of its surfaces indented
moulds for axe-heads of the simplest class.   In
this example there is no reason to believe that
any corresponding half was used to complete the
mould.   The melted metal was simply poured

into the indented surface, and left to take shape
by its equilibrium on the exposed surface.
Weapons formed in this way may frequently be
detected, while others, full of air-holes, and
roughly granulated on the surface, appear to have
been made in the still simpler mould formed by
an indentation in sand; others of the stone
moulds have consisted of pairs like those of
bronze. A very curious example of this description was found a few years since in the isle of
Anglesea. It is a cube of hone-stone, nine inches
and a-quarter in length, by four inches in breadth
at its widest extremity. Each of its four sides
are indented for casting different weapons—two
varieties of spear, a lance, or arrowhead, and a celt
with two loops. Only one stone was found, but
another, corresponding one is obviously requisite,
by means of which four complete moulds would be
obtained."

The means adopted for obtaining an intense
heat to melt and cast metals, were similar to ours
—by bellows or blowing through pipes with the
mouth. Wilkinson, in his "Ancient Egypt,"
has given the figure of a smelting operation.
The furnace seems only a heap of fire upon the
surface of the earth, and the bellows are two
large bags filled with air, upon which a man
is standing with a foot upon each bag—the
aperture of the bag being connected with a pipe
leading into the fire; while the man is seen put-

ting all his weight upon one bag to compress out the air into the fire, he is also seen lifting up his foot, and, at the same time, the upper fold of the other bag, by a string in his hand, by which the bag is being again filled with air. This apparatus is, no doubt, both simple and rude, and if it refers to the ordinary metallurgical operations performed by the nation, one could hardly suppose that castings of any great size could be obtained unless with great difficulty; nevertheless, such a simple apparatus may have been used by the Hebrews in the wilderness, both in the construction of the calf and the serpent. If the metal was melted in crucibles or pots, any number of these could be used to suit the work in hand. Wilkinson remarks, in reference to the bellows, " that rude as it may seem, it indicates the knowledge of certain fundamental principles, such as the valve, which had not been before accredited to them—for these bags, when emptied, could not be refilled and preserve their usefulness without a valve."

Concerning the quantity and cost of the various metals, we have no particular data; but it may be stated, that a metal, as well as any other object which has a known intrinsic use, and is required by a community, will rise and fall in value, according to the quantity obtainable and the cost of production; so that if a pound of iron is required, and it cost the value of the weight of an

ounce of gold to obtain it, and bring it to market, then that will be the relative value of the two metals. This law holds good in all ages and countries; and the fact of iron ore existing in abundance, will not lower the value of the metal if the cost of extracting it from the ore is great. For instance, the metal aluminium sells at a much higher price than iron, notwithstanding that the ore of that metal, which is common clay, is so abundant. In Scripture, iron is always referred to as an article of comparative low value, and of greater abundance than bronze, as is indicated by such passages as the following :—
"And gave for the service of the house of God of gold five thousand talents and ten thousand drams ; of silver, ten thousand talents ; of brass, eighteen thousand talents; and one hundred thousand talents of iron."

Hence, following these principles, we infer that the methods of extracting iron were so far simple and easy, enabling it to be sold cheaper than either tin or copper, but when we consider the articles recorded as having been made of iron, we think we have evidence sufficient that it was plentiful at a very early age. When iron and bronze are named together in certain proportions, we may form some general idea of the current quantity of one by the abundance of the other, especially when that other is the more valuable. The relative quantity of bronze and iron mentioned above is 18 bronze to 100 iron.

If anything like these proportions are maintained, and we believe they were greatly increased by time, the abundance of iron' must have been very great in the fourth century before Christ.

The perfection in the art of casting bronze attained by the Grecians, is amply borne out by the many specimens of such art still existing, and scattered throughout the various museums of Europe—a perfection which artificers of the present day endeavour to imitate—to surpass, is considered impossible. How far back we are to look to for the first attainment of this high state of perfection in the metallic arts is not known.

Rawlinson, speaking of the art of engraving as shown on the cylinder of Urukh, says—" The drawing on the cylinder was as good, and the engraving as well executed, as any work of the kind, either of the Assyrian or the late Babylonian period." There still exists on numerous jewels found in the ruins of ancient cities, evidence of the great skill possessed by Egyptians, Phœnicians, Greeks, and Romans, in the delicate art of engraving on precious stones, an art which gradually and greatly declined during the middle ages of the Christian era; and we have not yet been able to bring this art back to its former perfection, nor do we even know in many instances by what means these ancient workers produced such beautiful and intricate work. Our highest acquirements reach no further than a slavish imitation of workmen, who lived and died upwards of 3000 years ago, and whom we regard as belonging to the rude ages of the world.

It must have been observed by those who have read works on the genuineness or authenticity of the Pentateuch, and other historical books in the Old Testament, that one argument often used by a certain class of people is, that articles of manufacture are named in these books as being in common use at the time the books were written; while from profane history, it can be shown that the materials or substances capable of making such articles were not discovered till long after the reputed authors of the sacred books were dead, and, consequently, that such books could not have been written by the reputed authors, or the account of these articles must have been interpolated by later hands, and that in consequence Scripture history is not to be relied upon. We think the reader will notice the fallacy of this reasoning, which rests in this, that, especially in ancient times, certain arts have for periods been in common practice among a people or nation, but eventually, on account of war, or from other causes, the art at length ceased among them and knowledge of it became lost. We have an instance of this in the history of the Hebrews from the time they left

Egypt till David ascended the throne of the kingdom. Probably many other ancient nations, besides the Israelites, attained to great perfection in particular arts, but by a change of dynasty or a series of wars these arts fell into dissuetude and were lost. One instance of the above fallacious mode of argument may be stated. Because Plutarch and some other Grecian writers repeat a tradition that iron was not discovered till after the fall of Troy, it is argued that Joshua could not have been the writer of the book of Joshua, for in it is mentioned that the enemies of Israel used chariots of iron—"And all the Canaanites who dwelt in the valley have chariots of iron, both they who are of Bethshean and its towns, and they who are of the valley of Jezreel" (Joshua xvii. 16); and this was centuries before the fall of Troy. Notwithstanding the classic tradition of the Greeks, there is evidence of iron being in common use many ages before Joshua lived, and both Joshua and Moses in writing their histories spoke of things as they existed.

GOLD AND SILVER.

THE two metals, gold and silver, we intend to treat together, for the following reasons :—

I. They are almost always associated in nature.

II. In Scripture references they are generally both included.

III. The methods for their extraction and purification are similar.

Both gold and silver are found in nature, in the metallic state, in great abundance, and, consequently, these metals are considered to have been the first to attract the attention of man, more especially gold, that metal being widely diffused through the earth in small quantities, and in the locality of the primary or plutonic rocks, the quantity is sometimes. very large. The constant action of air and moisture upon the surface of these rocks causes their decomposition, and small portions are thus from time to time loosened and disintegrated. The gold in the rock is not subject to this decomposition, but is loosened by the decay of the stone, and carried from the high to the low levels by rain, where, washed by currents of water, the light earthy matters are carried away, and the gold left in a more concentrated state, consequently, in course of time, large tracts of country become covered, it may be many feet deep, with this washed debris of the mountain, composed of sand and particles of gold.   Hence, newly-discovered countries, where the primitive rocks exist, are often gold-producing, the production being limited to the extent of valley so covered, as well as the depth of the debris.   It often happens that, after the debris is exhausted, the original rock from which the valley had been filled is made productive by artificial grinding and washing, as is being done in California, Australia, and Mexico.   Whether or not this artificial washing of the ground, or pulverised rock, was practised in the earlier ages, we will have occasion to consider hereafter.   The natural condition in which gold is found in valleys, and on the margin of rivers, has very early notice in history.

Speaking of Eden, Moses says, "And a river went out of Eden to water the garden ; and from thence it was parted, and became into four heads. The name of the first is Pison : that is it which compasseth the whole land of Havilah, where there is gold ; and the gold of that land is good" (Gen. ii. 10–12).

And in the book of Job, which many able commentators consider to be as old as that of Moses, it is stated, "As for the earth, it has dust of gold."   The late Sir R. Murchison in his

"Ancient Siluria," remarks upon this passage in Job, as follows, " Modern science, instead of contradicting, only affirms the aphorism of the patriarch Job, who has shadowed forth the downward persistence of the ore (silver), and the superficial distribution of the other (gold); surely there is a vein for the silver, and the earth has dust of gold."

From these natural circumstances, so distinctly stated in the foregoing passages, taken in connection with the natural properties and character of gold, viz., its being always found in the metallic state, its high specific gravity, its colour, its polish, its malleability, &c., it is not surprising that it attracted the attention of the first inhabitants of the earth, and that they early began to use it for different purposes in the arts. In the first notice of gold, just quoted, there is a strong presumption that it had also been found in other localities than the one referred to, and that different qualities of the gold were known; hence the remark, " The gold of that land was good."

Gold is an element, and is not essentially good and bad, but in itself can only be of one quality; and, therefore, when different qualities are observed, it must depend upon certain admixtures. The only thing that can make gold bad is its having in combination some other metal or metals; so that the observation referred to in the text warrants us in concluding that, in that early age gold was an article of value, and that its value ranged according to its purity; but that they had methods for purifying the gold from the inferior metals alloyed with it, so early as the days Moses refers to, is not stated. That it was used for manufacturing articles, either for use or ornament, is highly probable, not only from the fact that it was familiarly known, but because iron and copper—two metals much more difficult to obtain and to work than gold—were used for some industrial purposes by the great-grandson of Cain, who, we are told, was an "instructer of every artificer in brass and iron" (Gen. iv. 22). Indeed, it is more than probable, that the working of brass and iron resulted from the experience they had obtained while endeavouring to work the gold, which we are led to believe was found in the locality where these artificers lived. If they knew how to make copper and iron from their ores, the gold could not possibly escape their notice.

We will now refer to the present methods for extracting the gold from the ores, or rather the earths, with which they are mixed; for, as already stated, gold is found in the metallic state in the primitive rocks or alluvial soil, in small thin leaves and threads, and little dendrical pieces attached to quartz, and sometimes in small loose particles in crevices of rocks, and also in pieces mixed with the alluvial soil, now popularly termed *nuggets*, · but more generally it is found in the form of small

grains like sand, mixed with the *débris* of the
primitive rocks. The metal is very seldom found
pure, but generally combined with silver, and
often copper. The gold from different localities is
characterised by the extent and kind of alloy which
it contains, and even the same locality will produce
little differences in quality; as for example, the
quality of Californian gold has been given by T.
H. Henry and Rivat in 100 parts :—

	No. 1.	No. 2.	No. 3.
Gold,	90·01	86·57	90·70
Silver,	9·01	12·33	8·80
Copper,	·86	·29	...
Iron,	...	·54	·38
	99·88	99·73	99·88

The following are the analyses of Australian
gold by A. D. Thomas, Australia :—

No. 1. Sample of metal coated with Iron-Oxide.

Gold,	87·78	87·77
Silver,	6·07	6·54
Oxide Iron,	6·15	5·69
	100·00	100·00

Freed entirely from Iron by fusion, gave—

Gold,	93·53	93·06
Silver,	6·47	6·94
	100·00	100·00

Another specimen of Gold gave—

Gold,	96·42
Silver,	3·58
	100·00

When the gold is found in large pieces, as in

nuggets, it can be nearly all separated from the earthy particles by mechanical means, then by fusion in a crucible, any remaining portions of earthy matter separate and float upon the surface.

When the grains of gold are small, and not easily separated from the earthy matrix by the hand, the whole is put into a crucible with a little borax, and fused; the gold is then obtained at the bottom of the crucible as a small button or ingot; but when the gold is found in the form of dust mixed up with the soil, whether naturally or formed by grinding the rock, the whole is first subjected to washing, either by hand-basins, or by those vessels and instruments termed cradles, by which means the greater portion of the earthy matters are washed away; or the dust is put into a current of water, which is made to pass at a certain velocity over a large flat surface; or, what is very common, over woollen cloths stretched out; the earth being lighter is carried along with the water current, and the gold falls and is retained by the cloth. This concentrated and washed gold dust is collected into proper vessels, and a quantity of mercury (quicksilver) is mixed with it, which dissolves or amalgamates with the gold or other metals, in the same manner as water dissolves sugar, and the earth is left upon the top of the mercury, and easily removed. In some fine gold dust sulphur and iron are present, and greatly interfere with the perfect reaction of the mercury; the dust is therefore, in such cases, submitted to a roasting which drives off the sulphur, and sets the gold entirely free to be taken up by the mercury. This mixture of mercury and gold, termed an amalgam, is subjected to a high heat in a retort, or what is more common, under an iron bell, when the mercury distills over and is recovered in the same manner as steam is condensed into water, and the gold and silver and other metals remain behind, and are separated by a process to be afterwards described.

Another method for extracting the pure gold from the gold dust, is by mixing with it a quantity of oxide of lead (*litharge*), and a little charcoal, with flux, if necessary, and bringing the whole to fusion in a crucible. The oxide of lead is reduced by the charcoal to the metallic state, and combines with the gold and silver present.

The lead, with the gold and silver, is put upon a hollow plate or flat vessel made of bone ashes, called a cupell, and submitted to a good red heat. When in this state, a current of air is blown upon the fused metals by bellows, which oxidises or scorifies the lead, and the gold and silver remain behind, pure but combined.

These are the methods now practised for obtaining the gold, or gold and silver, when in fine dust and mixed with other matters, as oxides of metals. The gold may also be procured from earthy matters, even although in fine dust, by

putting the mixture into a crucible or earthen pot,
and applying a strong heat; the gold melts and
collects at the bottom, and thus separates itself
from the earthy matters; and if lime, soda, or
such substances be added as a flux, these combine
with the earths, assist the fusion, and make the
whole operation easy and simple; although in this
process there is a much greater risk — nay,
almost a certainty—of loss. The metal being so
valuable, this method is not now adopted. Still
we are of opinion that this process was the one
practised in early ages for extracting the gold and
silver from their ores. All these processes refer
equally to ores of silver, but silver being often
found in a mineralised state, combined with other
substances, such as sulphur, chlorine, and oxygen,
its ores are generally subjected to a roasting or
calcining in a furnace before submitting them to
the amalgamation process. A very old method,
and one still extensively practised in Mexico, for
demineralising the silver, and causing it to combine
with the mercury, is that termed the Patio process.
The ore is finely ground and mixed with common
salt and sulphide of copper, previously roasted
and ground, and the whole made up into heaps,
and trodden out uniformly by mules. Mercury is
added from time to time, and in the course of
from twenty to twenty-five days the operation is
complete. The mercury is separated from the
mass by washing, and it then contains the silver

and gold which had existed in the ore. This amalgam is treated in the same manner as described above for gold by distillation.

Another method for mineralised silver ore is now adopted where opportunity presents itself. The ore is finely ground and mixed with common salt, and subjected to a dull red heat in a furnace; by this means the silver is converted into a chloride. The whole is then put into large casks or barrels with water, and scraps of iron and mercury. The iron combines with the chlorine, and the silver becomes reduced to the metallic state, and combines with the mercury, forming an amalgam, which is carefully collected at the termination of the process, and distilled as already described. If copper is in the ore, it forms part of the amalgam, and remains in the silver after the mercury is distilled off, and has to be separated afterwards either by the action of acid or cupellation. Although these processes are only applied to silver ore, yet any gold present is obtained along with the silver in the amalgam. In ores where the silver exists in the metallic state these processes are not necessary. When the lead process is applied, these last operations are also unnecessary —the raw ore being simply mixed with the litharge, charcoal, and flux, and fused, when the lead is obtained, containing all the silver and gold, which is afterwards cupelled, as in the case of gold.

The Indians in Peru perform their smelting operations in small portable furnaces or cylindrical tubes of clay pierced with holes. In these, they place layers of silver ore, galena, and charcoal, and the current of air which enters the holes quickens the heat and gives it a great degree of intensity. These furnaces are removed from one elevation to another to suit the force of the winds. . When the wind is strong, too much fuel is consumed. By this operation the natives obtain argentiferous masses, which they smelt again in their cottages by a species of cupellation. Ten or twelve persons sit round the fire, and through copper tubes from one to two yards in length, and contracted in the bore at one end, they blow upon it; by which means the lead is scorified and the silver obtained pure.

When gold and silver have copper or other inferior metals alloyed with them, which is often the case, they are purified by different methods, according to circumstances. The main principle involved in these methods may be thus stated:—

When gold or silver are subjected to a high heat, they remain, even during fusion, in a pure metallic state—that is, they do not become tarnished by combining with oxygen; on the contrary, if they somehow had got oxidised, a red heat would drive the oxygen away. If we take a piece of clean copper and make it red hot, it becomes black by combining with oxygen; if we take this coating of oxide off, and repeat the operation, another coating of oxide is formed; and so on until the whole copper is oxidised. The same takes place with tin and other inferior metals. If, then, we take an alloy of copper, gold, and silver, and expose it to a fusing heat in a current of air, the copper will thus combine with oxygen, and if the oxide is removed from the surface, the oxidation will continue as long as there is copper present,—leaving the gold and silver pure. If there be a large quantity of the alloy, the purification by this means will require a long time.

It has been long known that certain salts which contain large quantities of oxygen, when exposed to a high heat, give off this oxygen; one of these salts is *nitre* or *saltpetre*. This fact is, therefore, taken advantage of in purifying gold and silver. The alloy is fused in a vessel, such as a crucible, and, instead of exposing the fused alloy to the air, to oxidise the copper or other inferior metals, saltpetre is thrown upon the melted alloy, which, by giving its oxygen to the inferior metals, effects their oxidation, thus purifying the gold and silver. The impure alloy may also be mixed and fused with metallic lead, and then the whole subjected to cupellation.

Having thus described the operations for extracting and refining the precious metals, we have now, according to our plan, to inquire what references there are in Scripture or other ancient writings,

which will enable us to infer that any of these
operations were practised in ancient times? or, if
not any of these, what other? In this inquiry we
may often be startled with references which indicate greater knowledge of metallurgical operations
in ancient times than we were prepared to admit;
and if we begin at the earliest ages and come
down to the present time, we shall find evidence
that the advancements made in working the precious metals in modern times are not great in
extent. Pliny describes the amalgamation process
for extracting gold and silver from the earths,
and it is nearly the same as that practised at the
present day; and, going back in the inquiry, the
evidences go to prove that all metallurgical operations were not only practised, but the operators
and workers in metal were held in great esteem,
and constituted an important item in a nation's
prosperity; and that, during war, artificers in
metal were preserved and taken captive, in order
to enrich the conquering party, as is illustrated in
the following passage—"And all the men of might,
even seven thousand, and craftsmen and smiths a
thousand, even them the King of Babylon brought
captive to Babylon " (2 Kings xxiv. 16).

In ancient times, in Egypt, and doubtless also
in other countries, and until within the last two
centuries, an idea prevailed that gold was the only
true metal, and that all other metals were different
conditions of gold—that all metals could thus be

transmuted or converted into gold—and that some substance existed, or could be made, which would have the power of converting every metal into gold. This idea obtained for all metallurgical operations a high position, and gave to the investigations of the properties of the various metals, and their actions and reactions with one another, and with other substances, a prominent place in the studies of the learned. It need not, therefore, be wondered at that the ancients became skilled in many metallurgical processes, and also knew many substances capable of acting upon the baser, and separating these from the noble metals. So early as the days of Job, familiar mention is made of refining :— " Surely there is a mine for silver and a place for gold, which men refine " (Job xxviii. 1).

In Scripture, references to metallurgical operations are more frequent in the poetical than the historical portion. The poet, to illustrate or enforce some great moral fact or principle, descends for a time to some minute detail, in order to bring out the full force of the inference he intends to draw, and often a single word contains a deep and extensive meaning. A beautiful illustration of this is found in the following passage :—" The words of the Lord are pure words, even as silver tried in a furnace of earth purified many times " (Ps. xii. 6). Here we observe David is not satisfied with saying that the word of the Lord is pure as silver tried and purified many times: he wishes to represent the most perfect state of purity, consequently he descends to a particular method, and says, tried in a *furnace of earth*. This distinct reference to an earthen furnace leads naturally to the conclusion that different methods were then known and practised, and that the earthen furnace was considered best. Now, several methods have been described, by which we know gold and silver may be purified, independent of the process by mercury, to which we can find no reference in Scripture. These methods we will briefly repeat.

1st. Exposing the impure metal in fusion to a current of air. This method is not only tedious, but cannot be relied upon for perfectly purifying the gold and silver, even though often repeated.

2d. Keeping the alloy in a melted state, and throwing upon it nitre, with a little bone-ash or earthy matter, which aid the process. This is scummed off from time to time, and each time more nitre is added, until the precious metals are purified, which is known by the fused surface remaining bright. This process of purifying is quicker than the first, and gives pure metal.

3d. Mixing the impure metals with a quantity of lead, and subjecting the whole to fusion upon a flat vessel made as already described, either with bone-ash as in this country, or with clay lime and wood-ash as in Mexico and India, allowing no fuel to come into contact with the fused metals, but

blowing upon the fused mixture with bellows or
other means of causing a strong current of air, thus
oxidising all the inferior metals. The oxides thus
formed are partly absorbed by the cupell, which
is the bone-ash vessel, and are partly carried off by
the draft. This is by far the most perfect method
to ensure the complete purity of the gold and silver,
although the process may have to be repeated.

Knowing that the cupelling process, No. 3., was
practised at a very early period, we think it probable that it is the method of purification referred
to by the Psalmist. That the purification of the
precious metals was performed by the cupelling
process, very shortly after David's time, is clearly
indicated in various parts of Scripture. Indeed,
some of the early prophets, as well as those of a
later date, give a glowing description of the whole
process, and draw from it a spiritual lesson.
"Thy silver is become dross—I will bring again
my hand upon thee, I will purify thy dross and
take away all thy tin " (alloy) (Isa. i. 25) ; according to the marginal reading it is, " I will take
away thy dross in the furnace." The prophet
Jeremiah is more explicit. " They are all brass
and iron ; they are all corrupters. The bellows
are burned, the lead is consumed of the fire ; the
founder melteth in vain : for the wicked are not
plucked away. Reprobate silver shall men call
them" (Jer. vi. 28–30). This description is very
perfect. If we take silver having the impurities

in it referred to in the text, namely, iron, copper, and tin, and mix it with' lead, and place it in the fire upon a cupell, it soon melts; the lead will oxidise, and form a thick, coarse crust upon the surface, and thus consume away, but effect no purifying influence. The alloy remains, if anything, worse than before.

The prophet says that the alloy was put into the fire with lead—the whole was melted. There is no lack of heat; but the lead was consumed—that which should have purified the silver passed away as dross, and the silver was left unrefined, because "the bellows were burned"—there existed nothing to blow upon it. Lead is the purifier, but only so in connection with a blast of air blown upon the melted metals. We cannot pass over the aptness of the illustration in its application to affliction being sent upon a sinful people, to draw them from their wickedness, represented by the heat and melting, and how ineffectual these will be in removing the sin, represented by the impure metals, if the Spirit do not accompany the operation, and blow upon it.

Ezekiel, using the same process as a figure, gives the position of the cupell in the furnace. "Son of man, the house of Israel is to me become as dross: all they are brass, and tin, and iron, and lead, in the midst of the furnace; they are even the dross of silver. Therefore, thus saith the Lord God, Because ye are all become dross, behold, therefore I will gather you in the midst of Jerusalem. As they gather silver, and brass, and iron, and lead, and tin, into the midst of the furnace, to blow the fire upon it, to melt it; so will I gather you in mine anger and in my fury, and I will leave you there, and melt you. Yea, I will gather you, and *blow upon you in the fire* of my wrath, and ye shall be melted in the midst thereof. As silver is melted in the midst of the furnace, so shall ye be melted in the midst thereof" (Ezek. xxii. 18–22). This is a very graphic description of the process of cupellation. Collecting the impure silver and mixing it with lead, and putting the whole in the midst of the furnace, no doubt upon a cupell, and when melted, blowing upon it for the purpose of scorifying and burning out all the impurities. The different commentators we have consulted upon these verses, refer to the materials being collected and put into the furnace, simply for the purpose of being melted; the blowing having reference to the fire in order to produce an intense heat, such as in blast or cupola furnaces. Now, we think these explanations are erroneous and inapplicable to the circumstances, and they entirely destroy the beauty of the figure. For, such an alloy as that named in the text melts at a very low heat—not more than can be produced in a common kitchen fire—therefore the blowing of the fire to get intensity of heat is not necessary.

Again, were the materials named in the text
put into the furnace amongst the fuel, and the
fire blown as is done in a blast or cupola furnace,
the results which are said to follow could not
occur, as the presence of the fuel with the metal
would prevent scorification, no matter how intense
the heat.

Now, compare the text with the cupelling process. The vessel or cupell containing the alloy is
surrounded by the fire, or, in the words of the
prophet, placed in the midst of it, and the blowing, as will be observed, is not directed on the fire,
but on the fused metals—figuratively, Israel—the
object to be purified; and when this view of the
matter is taken, the figure is perfect, as nothing
can resist this scorifying process.   Intense heat
is not required for scorification.

There is another appropriate reference to these
operations, which still further proves the correctness of our views on their refining operations with
gold and silver: "Who shall stand when He
appeareth? for He is like a refiner's fire: and He
shall sit as a refiner and purifier of silver, and He
shall purify the sons of Levi, and cleanse them as
gold and silver" (Mal. iii. 2, 3).   There is
nothing in this passage directly referring to the
process of refining, further than the position of the
refiner and the result of the operation.   We will,
however, endeavour to illustrate this reference.
When the alloy is melted as before described,

upon a cupell, and the air blown upon it, the surface of the melted metals assumes a deep orange
red colour, and a kind of flickering wave of colour
constantly passes over the surface, caused by the
vapouring of the oxygenated impurities. As the
process proceeds, the heat is increased, because
the nearer the metals approach purity, the more
heat is necessary to keep them in fusion. After
a time the colour of the fused metals becomes
lighter—the impurities only forming reddish striæ,
which continue to flit over the surface. At this
stage, the refiner watches the operation, with
great earnestness, until all the orange colour
and striæ disappear, and the metal assumes a
bright mirror-like appearance, reflecting every
object around; and the refiner, as he looks upon
the mass of refined metal, may see his own
image reflected therefrom, and thus he can form
a very correct judgment respecting the purity of
the metal. If he is satisfied, the fire is withdrawn and the metal removed from the furnace;
but if the metal does not assume this mirror-
like appearance, more lead is added and the
process continued.

This text may also apply to a refiner at the
side of the furnace, casting into the melting
pot the scorifying flux, such as nitre, to burn
away the dross, and watching its success. The
metal in this case also assumes that mirror-like

appearance described, but the cupelling operation being well known in the days of Amos, and being most suitable to the figure of the text, we are inclined to believe that the sacred poet refers to the cupelling operation.

The following are a few more extracts from Scripture referring to the purification of gold and silver.

"When he hath tried me, I shall come forth as gold" (Job. xxiii. 9).

"Thou hast tried us, as silver is tried" (Ps. lxvi. 10).

"Take away the dross from the silver, and there shall come forth a vessel for the finer" (Prov. xxv. 4).

"I will bring the third part through the fire, and will refine them as silver is refined, and will try them as gold is tried" (Zech. xiii. 9).

These statements and warnings addressed to the people, evidence that the operations referred to were well known, otherwise their application would not have been understood.

We have repeatedly stated that all these operations are equally applicable to the purification of gold as well as silver, so that when they are associated in the raw ore, as they generally are, they will be found together at the termination of the refining process.

Gold is seldom, if ever, found in nature free from

silver, the quantity of silver varying from 2 to
30 per cent. Silver again is seldom found free
from gold, the quantity of gold varying from 1 to 5
per cent. The separation of these two metals from
each other is an operation performed after the
cupelling or purifying we have described is completed. This operation is termed, in modern
technical phraseology, PARTING, and is performed
in the following manner :—

The alloy is beat into thin plates, and the silver
separated by dissolving it out, with either nitric or
sulphuric acid. If gold prevails in the alloy, more
silver has to be added, to render it better fit for
the parting process. When nitric acid is used as
the solvent, the proportions of the metals should
be three of silver to one of gold; and when
sulphuric acid is used, six of the former to one of
the latter. On boiling the thin plates of the alloy
in the acid the silver only is dissolved, and the
gold remains, the solution of silver is subsequently
poured off, and the gold, after being thoroughly
washed, is melted and cast into ingots or bars.
To the solution of silver a quantity of common
salt (chloride of sodium) is added, which pre-
cipitates the silver as a chloride; this is collected
and fused with carbonate of soda or potash, the
silver being then obtained as a pure metal; or
there may be placed in the solution of silver,
sheets of copper, when the copper combines with
the acid, and the silver is deposited as metal, and

is then collected, washed, fused, and cast into
ingots.

We cannot find any reference in Scripture, or
other ancient writings, which enable us to infer
whether or not these operations for parting small
quantities of silver from gold were practised in
ancient times. Only two passages occur that
refer to distinct and different operations being
applied to gold and silver. These are—" The
fining-pot is for silver, and the furnace for gold;"
and " As the fining-pot for silver, and the furnace
for gold " (Prov. xvii. 3. and xxvii. 21). These are
translated by some commentators—" As silver is
tried by the fining-pot and gold by the furnace."
Commentators also generally state that the *fining-
pot* means crucible, or melting-pot. If this be
the case, we know of no operation practised in
modern times that the above passages will refer to
—in other words, there is no operation in which
a melting-pot or crucible is used that is peculiar
to silver, nor any peculiar to gold that is performed
in a furnace. All such operations apply equally
to both; and if any distinct operation existed at
the time of Solomon, the process, so far as we are
aware, is now lost. It may be, however, that the
word fining-pot does not refer to a crucible, but
to some operation practised for parting. If, for
example, the term *fining pot* could refer to the
vessel or pot in which the silver is dissolved from
the gold in the parting process, as it may be called

with propriety, then these passages have a meaning in accordance with our modern practice, and suit admirably the inference Solomon draws—a trying or testing of a man's character. However, this is but a supposition, and must be taken as such, even although plausible, for they may have had means of testing the quality of gold and silver, and separating them, in the days of Solomon that we have not. Excepting the above passages, there is no other reference in Scripture to operations applied to gold and silver separately; but with regard to the substances used in the operation of parting, it is well known that the ancient Egyptians were acquainted with both nitric and sulphuric acids, and also knew that these acids dissolved silver. The writings and markings upon ancient mummy cloth are now found to have been done by a solution of silver—and the solution of silver which they used is all but positively ascertained to have been the nitrate, the marking ink of our own day, which is strong presumptive evidence as to their knowledge of the action of acids upon silver; and this is also a circumstantial evidence in favour of their using such means for separating silver from gold, as the fact could hardly escape their notice, for in dissolving the silver the gold would be left. But to assert that they were acquainted with all the requirements now known to us for obtaining perfectly pure gold and silver by such operations, would be more than our references warrant; in-

deed, the repeated references to certain qualities
and kinds of gold in Scripture, is a kind of presumptive proof that they were not in the habit
of perfectly purifying or separating the gold from
the silver, and the small quantities of silver remaining would give a distinctive character to the
gold. If the parting process was not perfectly
understood, the silver would not be perfectly
separated, which would produce certain distinctive
qualities of gold of less or more value. Different
localities were known to produce gold of a certain
fineness, either from its being in a pure state
natively, or imported from these localities after
being subjected to the operations of parting; these
qualities were named after the localities from which
they were brought, as " the gold of *Opher*," " gold
of Parvaim," " gold from the north," &c., &c.
Parvaim is said to mean eastern, and may be in
contradistinction to gold from the north.

Mr Charles Beke, in the *Athenæum* of date Nov.
1868, says that "Opher was in Persia; the gold field
was in Eastern Africa, the shipping port being at
the Gulf of Persia (the land of Opher); the gold
was named after the place it was shipped from,
the same as many other articles are at the present
day, such as Mocha coffee, Turkey rhubarb, &c.,
named from the place they are imported from, but
not where they grow. The voyage mentioned in
the Bible, from Ezion Geber to Opher, is said to
be three years, but it may have been less than

two years, as by Hebrew rule any part of a year
was called a year; thus, say the voyage began in
December, and ended in January following, it would
be called two years; and if it took till January
twelvemonths, it would be called three years,
although in reality only fourteen months; so that
the voyage there and back may have occupied in
reality only two years." Beckmann, in his "History
of Inventions," states positively that the ancients
did not know the means of separating gold and
silver: "The ancients used as a peculiar metal a
mixture of gold and silver, because they were not
acquainted with the art of separating them, and
afterwards gave it the name of *electrum.*" He
thinks they considered this a distinct metal, but
Pliny mentions electrum to have been made
artificially by mixing four parts gold and one of
silver, proving that in his day they did not consider
it a distinct metal.

We think, from what has been shown in the
references produced from Scripture upon the
metallurgical operations of gold and silver, that
the ancients were well versed in these arts, and
that many of the most delicate operations, requiring
considerable chemical skill, were practised by them.
The advancements made in these metallurgical
operations in modern times have not been so new
to the world as we are sometimes led to suppose.
However, we will have opportunities of illustrating, in other parts of this volume, how it

happens that advancement in the arts has not
been a steady progression from early days to the ·
present; but that processes and operations well
known to the ancients have become lost, and after
a lapse of time, been again discovered, requiring
many years to regain their former perfection.   So
that, taking our stand at the present date, and
looking back to the palmy days of Egypt, Greece,
or Israel, we know little of what has passed from
us in the interval; and except from the survivals
of ancient art, or where the poet, to adorn his
verse, has borrowed from the arts an analogy to
illustrate passing life, we have little knowledge of
how the ancient world did in the practical operations of metallurgy.

A list of gold and silver articles in common use
in ancient times, would embrace all the articles
and purposes to which these metals are applied
at the present time, both for domestic and public
use, indicating extensive manufacturing operations; and even the delicate and fine operations of
gilding and plating wood and inferior metals with
gold and silver, were also practised in ancient
times.   Sheet silver is referred to in Scripture,
"Silver spread into plates is brought from
Tarshish, and gold from Uphaz, the work of the
workman, and of the hands of the founder; blue
and purple is their clothing : they are all the work
of cunning men."   Had they rolling mills in
those days, or were the sheets beat out with the

hammer? We think it was the work of the hammer, and the process is referred to as being employed at an earlier age than that of the quotation, both for forming plates of gold and silver. Gold and silver, beat out thin, and cut into threads for sewing or weaving, are referred to in the construction of the ark ; and Wilkinson states that gold was beat out in leaf, and consequently into plates, in ancient Egypt.

Gold and silver were used in ancient times for the purposes of exchange, either in the form of coined money, or some other forms of a standard quality, and weighed when exchange was made.

In Siam at the present day the currency is silver, and no payments are made in minted pieces of gold. If in large transactions gold is used, it is in the form of leaf or bars, and is weighed. We read in Scripture that Abraham bought the cave and field of Machpelah from Ephron, with silver weighed over to him in the current money of the time. "And Abraham weighed to Ephron the silver which he had named in the audience of the sons of Heth, four hundred shekels of silver, current money with the merchants."

Some commentators argue that because such a sum is named in silver, gold could not have been used for coinage or exchange at that time ; this is a negative way of reasoning. The French, for example, name large sums of money in francs— silver coins ; but it would be a great mistake to

conclude from this that the French had no gold
coinage. Or supposing that but one transaction had
been recorded in the reign of one of our early kings,
such as one hundred and twenty shillings having
been paid for something or another, we would not
think of reasoning therefrom that gold was not
then in use for the purposes of exchange; the fact
of silver being so distinctly named in the text, in
itself favours the opinion that other metals were
in use for exchange purposes.

It is also stated that because the silver was
weighed it could not have been in the form of
coined money, but pieces of silver probably stamped
to indicate the quality; we think that the passage
refers to pieces of silver of a fixed weight, and
used as coins whatever may have been their shape,
and not to the mere weighing of a heterogenous
lot of silver. The word shekel is a name for a
coin as well as for a weight. At the present day
in some parts of the east where regular coins are
used, but owing to their not being milled on the
edges, or other precautions taken to prevent them
from being scraped and lightened, the just and
honourable merchant in transferring a number of
them weighs them. The practice may have been
similar in ancient times, but Abraham, who, besides being an upright man in all his transactions,
knew well the character of the people he had to do
with, may therefore have weighed out the four
hundred shekels of money. So that the weigh-

ing was an act of policy and strict commercial honesty.

Thomson, in " The Land and the Book," has the following remarks upon this transaction between Abraham and Ephron :—" My experience in these transactions leads me to believe that this price was treble the actual value of the field. But, says the courteous Hittite, four hundred shekels, what is that between thee and me ? Oh how often you hear these identical words on similar occasions! and yet, acting on their apparent import, you would soon find out what and how much they meant. Abraham knew that too, and as he was then in no humour to chaffer with the owner, whatever might be his price, he proceeded forthwith to weigh out the money. Even this is still common ; for although coins have now a definite name, size, and value, yet every merchant carries a small apparatus by which he weighs each coin, to see that it has not been tampered with by Jewish clippers." And he further says that the deeds of such a purchase are the same now as then : every item must be enumerated. " The cave that was therein, and all the trees that were in the field, and that were in the borders round about, were made sure."

The fact of their having weights for transfers, is of itself a proof of considerable advancement in commercial dealings, as a fixed standard of weight must previously have been agreed upon. Probably the balance was first used for the purpose of weighing metals or money as articles of exchange. This transaction of Abraham's suggests another necessity, namely, a standard of quality in the metal, whether coined or otherwise. It has been shown that the value of the precious metals varies according to their purity; fifty shekels, by weight, of one quality, may not be worth more than the half of fifty of another quality; therefore some fixed standard of quality must have been agreed on in commerce to insure justice and equality in all money transactions. The passage does not say that Abraham gave four hundred shekels of refined silver, but four hundred shekels of silver current money with the merchants. This was, no doubt, money of a fixed standard of value agreed to, or understood, by the merchants; and the possession of such a standard implies not only a knowledge of the refining of the metal to bring it to the required purity, but also a method of testing, whereby silver, not up to the standard, could be detected, and not allowed to pass as current money with the merchants.

The quantity of the precious metals possessed by men in ancient times is another matter which has startled historians and commentators from its vastness, and has caused many to question the accuracy of the statements, because we have nothing equal to it now; a style of argument which is very erroneous, as it takes our own experience, or the experience of the present, as the proper test whereby to judge of the past, and this often without duly considering all the circumstances of the two periods.

For several centuries previous to the last forty years, gold and silver were very scarce throughout the world. The communities through which they have become diffused are very extensive and widespread, so that great accumulations in one place have not been permitted, and are hardly possible in our commercial age. In ancient times this diffusive influence was much less, hence greater accumulations were made of the precious metals, both by individuals and nations. Besides, there have been periods in the world's history when gold and silver were got in very great abundance compared with other periods; and the period from which we intend to take our data for comparison with the present age was probably one of the most favourable recorded in the history of the world, namely, the days of Solomon.

According to estimates made by Baron Humboldt, the whole produce of the precious metals from all the known countries in the world, from 1800 to 1848, was in value about £390,000,000. Since 1848, there have been discovered the new gold fields of California and Australia. By another authority, the total amount of gold in use in 1848 was estimated at £600,000,000, and the annual supply was believed to be between eight

million and nine million sterling, but since that
time the annual supply has been greatly on the
increase.   From a French work by M. Roswag—
" Metaux Precieux "—we extract the following
estimate of the amount of the precious metals
in the world in 1857, given in sterling money :—

Silver value	£1,347,125,000
Gold „	851,250,000
	£2,198,375,000

He calculated the average yearly increase at the
rate of—

Silver value	£10,000,000
Gold „	20,833,333
	£30,833,333

Taking this data from 1857 to 1872=fifteen
years, we have—

Silver value	£150,000,000
Gold „	312,499,995
	£462,499,995

which, added to that of 1857, gives as the gross
quantity of the precious metals in the world at the
end of 1872, not deducting tear and wear—

Silver value	£1,497,125,000
Gold „	1,163,749,995
	£2,660,874,995

In going back to Biblical times, we have no such
data as the statistics of mines throughout the
whole world; but at the period we have referred

to, we have an account of an extraordinary and systematic accumulation of the precious metals for a specific object, stretching over forty years. From the day David was anointed king of Israel, he seemed to have been fully impressed with the belief in God's promise, that he was to be the instrument for establishing the Israelites as the first nation in the world, in connection with the worship of the true God; hence every victory he achieved was looked upon by him as the means which God was using for effecting this great end; and after defraying the necessary expenses, remunerating and rewarding his people and army, the remainder of the spoils taken in war were accumulated for the great object of his life, as is shown in such passages as the following: "And David took the shields of gold that were on the servants of Hadadezer, and brought them to Jerusalem. And from Betah, and from Berothai, cities of Hadadezer, king David took exceeding much brass. . . . And Joram brought with him vessels of silver, and vessels of gold, and vessels of brass: which also king David did dedicate unto the Lord, with the silver and gold that he had dedicated of all nations which he subdued" (2 Sam. viii. 7–11). "And he took their king's crown from off his head, the weight whereof was a talent of gold, with the precious stones: and it was set on David's head. And he brought forth the spoil of the city in great abundance" (2 Sam. xii. 30). Even before

David's reign this setting aside for God a part of
all the spoils taken in war, was practised; and
much of what had been dedicated was in existence
at this time, as is indicated by such passages as
the following: "And all that Samuel the seer,
and Saul the son of Kish, and Abner the son of
Ner, and Joab the son of Zeruiah, had dedicated,
and whosoever had dedicated anything, it was
under the hand of Shelomith and of his brethren."
But none, evidently, had the same fixed object in
view as David, which object he made the prominent aim of his life. The law of Moses forbade
kings accumulating gold and silver for private
purposes: "Neither shall he greatly multiply to
himself silver and gold" (Deut. xvii. 17). This
probably accounts for David's extraordinary personal liberality to the cause, and is another proof
of the sincerity with which he acted in all religious
matters.

No sooner had David subdued the enemies of
Israel and established his kingdom, than he began
to develop this great object of his life—the erection
of a temple—to establish religious worship in his
kingdom, and make Jerusalem its great centre.
He first calls an assembly of the priests, the nobles,
and elders of the people, representing the whole
interests of the nation, along with Solomon, his
successor, and divulges to them his great plan,
the instructions he had received from on high, and
what he wished them to do in helping on this

great work. He also specifies his accumulations
for this purpose : " Behold in my trouble I have
prepared for the house of the Lord, a hundred
thousand talents of gold, and a thousand thousand
talents of silver, and of brass and iron without
weight; for it is in abundance " (1 Chron. xxii.
14). David, finding that his days were rapidly
shortening, while his anxieties increased for the
peace and welfare of his young successor, and for
the fulfilment of his great work, calls another and
probably more numerous meeting of the representatives of the people, and earnestly urges their
interest in the cause in the following language :
" Now I have prepared with all my might for the
house of my God, the gold for things to be made
of gold, the silver for things to be made of silver,
and the brass for things to be made of brass, and
iron for things to be made of iron, and wood for
things to be made of wood, onyx stones, and
stones to be set, glittering stones of divers colours,
and all manner of precious stones and marble
stones in abundance. Moreover, because I have
set my affections to the house of my God, I have
of mine own proper goods of gold and silver,
which I have given to the house of my God, over
and above all that I have prepared for the holy
house, even three thousand talents of gold, of the
gold of Ophir, and seven thousand talents of
refined silver to overlay the wall of the house
with " (1 Chron. xxix. 2-4).

The response to this appeal and example was very satisfactory, and is no doubt the largest voluntary collection ever made for church building purposes: "Then the chief of the fathers and princes of the tribes of Israel, and the captains of thousands and of hundreds, with the rulers of the king's work, offered willingly, and gave for the service of the house of God of gold five thousand talents and ten thousand drams, and of silver ten thousand talents, and of brass eighteen thousand talents, and a hundred thousand talents of iron, besides precious stones," &c. (1 Chron. xxix. 6, 7).

This contribution of David from his personal property is supposed by some to be included in the sum mentioned in our first extract; but this we think erroneous. Not only is this a separate sum, but it is stated by the donor that it is to be applied for a specific purpose—the overlaying of the walls of the house—and its quality is referred to as being the best, "the gold of Ophir, and refined silver," and it is also evident from the following passage that the others had already contributed from their share of the spoils to make up the large sum accumulated: "Which Shelomith and his brethren were over all the treasures of the dedicated things, which David the king, and the chief fathers, the captains over thousands and the captains over hundreds, and the captains of the host, had dedicated. Out of the spoils won in battle did they dedicate to maintain the

house of the Lord" (1 Chron. xxvi. 26, 27).
So that the voluntary contributions made by David
and the representatives of the people were extra
contributions. The gross sums calculated and
reduced to our standard for the sake of comparison, stand as follows:—

Talent of gold worth £5475; the talent being
125 lbs. in weight, will make the value of the
gold about 73s. per ounce. The talent of silver
is valued at £342, 3s. 9d., or 4s. 4½d. per ounce.
The gross of the value given will be—

Sum accumulated and in the public Treasury,	Gold,	£547,500,000
	Silver,	342,187,500
Contributed by David from his private resources,	Gold,	16,425,000
	Silver,	2,395,312
Contributed by the people,	Gold,	28,000,000
	Silver,	3,421,875
		£939,929,687
Total value of Gold,		£591,925,000
„ „ Silver,		348,004,687

When we consider the weight of all this metal,
amounting to many thousand tons, it is not surprising that many are startled, and are rather
more inclined to suppose that errors have occurred
in the transcribing, than that so vast an accumulation could have been made by any one people.

But the quantity of the precious metals possessed by the Israelites at that period is no crite-
rion of what was in use in the whole world. The

Phœnicians were then the masters of the sea, and had extensive commercial relations with most known countries, and had in their possession a vast number of miners working in different mines. Solomon was wise enough to perceive that to maintain the greatness of Israel in peace, they must become a commercial people. He therefore made arrangements with Hiram, king of Tyre, for ships and men to navigate them, to instruct his people in this art, and he also fitted out a navy for himself, as is stated: " And Hiram sent him by the hands of his servants' ships, and servants that had knowledge of the sea " (2 Chron. viii. 18). " And king Solomon made a navy of ships in Ezion-geber, which is beside Eloth, on the shore of the Red Sea, in the land of Edom. And Hiram sent in the navy his servants, shipmen that had knowledge of the sea, with the servants of Solomon. And they came to Ophir, and fetched from thence gold, four hundred and twenty talents, and brought it to king Solomon " (1 Kings ix. 26-28). " And again, it is stated that " The weight of gold that came to Solomon in one year was six hundred and threescore and six talents of gold ; beside that which chapmen and merchants brought. And all the kings of Arabia and governors of the country brought gold and silver to Solomon " (2 Chron. ix. 13, 14).

Lawson, in his Scripture Gazetteer says :—" It is traditionally related that when the Phœnicians

visited Spain, they found the silver in such abundance, that they not only loaded their ships to the
water's edge, but made their common utensils and
even anchors of this metal. This statement harmonises with the representations given by the
Spanish discoverers of Peru; and, whether exaggerated or not, certain it is that the Phœnicians
lost no time in taking possession of the country,
and forming colonies in the present Andalusia."

"When the Phœnicians first settled in Iberia,
artificial mine works were unnecessary; the ore
lay exposed to view, and a light incision was all
that was requisite to procure it in abundance.
The inhabitants were little acquainted with its
importance until the demands of the commercial
adventurers, and their avidity to possess it, first
taught them its value. When the stock which the
inhabitants had on hand, and for which they received various articles in exchange, was exhausted,
the Phœnicians saw it necessary to open mines,
and the fate of the Iberians afterwards became
deplorable."

If the period to which this tradition refers be
near the time of Solomon, it harmonises with the
statement made in Scripture, "Silver being little
thought of in those days, being plentiful as
stones."

Aristotle refers to one visit to Tartessus where
for the oil and other products of little value which
they had in their vessels, they received so much

silver, that they were unable to carry it, and at
last cut off the masses of lead which served them
as anchors, and substituted silver in the place of
them.   This is important proof, not only' of the
abundance of silver, but also of lead, seeing that
masses of it were used for anchors.

There are other facts referred to in Scripture,
which prove that the precious metals were much
more abundant in ancient than modern times.
The queen of Sheba, along with other gifts, brought
gold with her from her own kingdom to Solomon:
" And she gave the king one hundred and twenty
talents of gold " (2 Chron ix. 9) — equal to
£657,000.

When Haman, the favourite of the Persian king,
Ahasuerus, wished to prosecute his ambitious
designs, and proposed his plan for destroying the
Jews, thinking that it might be objected to from
the effect it was likely to have upon the national
exchequer, he makes the following offer: " If it
please the king, let it be written that they may
be destroyed: and I will pay ten thousand talents
of silver to the hands of those that have the
charge of the business, to bring it into the king's
treasuries " (Esther iii. 9).   If the talent referred
to be the Babylonish talent, the sum will be about
£2,119,000; but if reckoned by the Jewish talent,
it will be upwards of £3,000,000, offered by a
private person for an object of ambition—a sum
which would not affect his position as a man of

wealth, or it would have frustrated his object. Gold is not named by Haman; but we are not to suppose that this whole wealth in precious metals was in silver, as we know from history that gold was very abundant amongst the Persians at that time. Herodotus relates that when Xerxes went into Greece, Pythius, the Lydian, had 2000 talents of silver, and 4,000,000 of gold darics, which is estimated at £5,500,000.

We were not a little surprised to find Mr Gladstone, in his " Juventus Mundi," say that in Homer's time, silver seems to have been much more rare than gold, because it was chiefly obtained by scientific means. David and Homer were nearly contemporaneous authors, and the relative quantity of silver provided and given for the erection and furnishing of the Temple is about nine times that of gold, so that silver could not have been so very rare.

We take the following, from Herodotus, in reference to the temple of Belus, from Kitto's "Cyclopædia of Biblical Literature:" — "And there belongs to the temple in Babylon another shrine lower down, where there stands a large golden image of the god, and near it is placed a large golden table, and the pedestal and throne are gold; and, as the Chaldeans say, these things were made for eight hundred talents of gold, and out of the shrine is a golden altar. And there is another great altar where sheep offerings are sacrificed, for it is not permitted to sacrifice upon the golden altar except sucklings only; but upon the greater altar, the Chaldeans offer every year a thousand talents' worth of frankincense at the time when they celebrate the festival of the god. And there was at that time a statue of twelve cubits of solid gold; but I did not see it, but relate merely what was told to me by the Chaldeans. Darius Hystaspis wished to have this statue, but did not dare to take it; but Xerxes, his son, took it, and slew the priest who forbade him to move the statue. Thus is the sacred place adorned, and there is also in it many private offerings. These offerings made by individuals, consisting of statues, censers, cups, and sacred vessels of massy gold, constituted a property of immense value. On the top, Semiramis placed three golden statues of Jupiter, Juno, and Rhea. The first was forty feet high, and weighed one thousand talents. The statue of Rhea was of the same weight: the goddess was seated on a golden throne with lions at each knee, and two serpents of silver. The statue of Juno was erect like that of Jupiter, weighing eight hundred talents: she grasped a serpent by the head with her right hand, and held in her left a sceptre enriched with gems. A table of beaten gold was common to these three divinities, weighing five hundred talents. On the table were two goblets of thirty talents, and two censers of five hundred talents

each, and three vases of prodigious magnitude.
The total value of the precious articles and
treasures contained in this proud achievement of
idolatry has been computed to exceed one hundred
and twenty millions sterling." The amount of
the gold and silver taken by Cyrus when he
conquered Asia, according to the account of Pliny,
was £126,224,000 of our money.

We may mention here what is related of Sarda-
napolis: " When his last hope was extinguished of
saving the city, he fled into his palace, and order-
ing a vast pile to be reared in the court, on which
he accumulated all his treasure, amounting to a
prodigious value, and close to which he placed his
eunuchs, his concubines, and lastly himself, he
set fire to it, and perished amidst the splendid
ruins. Athenians represent these treasures as
worth a thousand myriads of talents of gold, and
ten times as many talents of silver—this is fourteen hundred millions sterling."

Pliny mentions that Pompeius Magnus, after
his third triumph over the kings and nations of
Asia made the following presents, " To the state
he presented two thousand millions of sesterces ;
to the legati and quæstors who had exerted
themselves in defence of the sea coast he gave one
thousand millions of sesterces, and to each individual soldier six thousand sesterces," the number
of soldiers who received this sum is not given.

The sum given to the state is equal to £16,666,666
                To the legati   .   .   8,333,333
                                      ─────────────
                                       £24,999,999

Mr G. Rawlinson, writing about the wealth of
the palace of Susan or Persepolis, says, "The golden
throne of the monarch stood under the embroidered canopy or awning supported by pillars of
gold inlaid with precious stones; couches resplendent with silver and gold filled the rooms." And in
Ecbatana, the palace, which, according to a writer
in the "Encyclopædia Metropolitana," "was nearly
a mile in circumference, the roof was covered with
tiles of silver, and the beams, ceilings, and pillars
were all covered with scales of silver and gold."

Gahn, vol. ii., states that "Crassus, proconsul
of Rome, entered Jerusalem to pillage the Temple.
Eleazer, the treasurer of the Temple, promised him
a bar of gold, weighing 300 mina, which was
preserved in a beam at the entrance to the Holy
of Holies, on condition that he would leave the
remainder of the treasures untouched. This condition Crassus solemnly swore to observe, but as
soon as he had obtained the golden bar, he robbed
the temple of two thousand talents which Pompey
had left, and took eight thousand talents besides."
Taking these sums by Jewish weights it would
equal £54,782,850. Josephus says, that when
Jerusalem was taken, the quantity of gold was
so great, that in Syria the gold fell to one half

its value." The Jews were always and still are
the greatest accumulators of the precious metals.
From these and many other statements in history,
we think we have conclusive evidence that gold
and silver were much more plentiful in ancient
than they are in more modern times; at all
events they were accumulated in greater quantities
in certain localities than they now are; and therefore we do not see why we should doubt the statements of Scripture history that such a quantity
of precious metals were in the hands of the
Israelites.

In the meantime, and before we leave for the
present the consideration of gold and silver, we
will offer a few observations upon that puzzle to
commentators, and object of ridicule to infidels
and others, namely, the destruction of the golden
calf made by Aaron in the wilderness, which in
its manufacture and destruction is one of the most
interesting circumstances in the Scripture history
of the metallic arts. For the sake of our non-
practical readers we may explain that from the
high malleable property of gold, it cannot be
ground. If a piece of gold be taken and pounded
in a mortar, or hammered, it will get flattened to
almost any extent, but will never be ground to
powder. Now as the Scripture says that Moses
ground the calf to a powder, the question arises,
In what way was it done? Some commentators
are of opinion that the calf was not ground, that

Moses deceived the people, keeping the gold, and substituting something else instead, which he made them believe was the gold. Others, again, unwilling to admit the above supposition, or confess that they are ignorant of the method, have mentioned other means as being more probable, and thus get over the difficulty to their own satisfaction. Finding it stated in some old book on chemistry, that gold could be made potable or soluble by a certain mixture, this has been taken hold of as the probable if not the veritable process adopted by Moses. A few instances will suffice to show how much has been made of a mere figment.

Our first quotation is from Dr Kitto's "History of Palestine:" "Commentators have been much perplexed to explain how Moses burned the golden image and reduced it to powder. Most of them offer only vain and improbable conjectures, but an able chemist has removed every difficulty on the subject, and has suggested this simple process as that which Moses employed : Instead of tartaric acid, which we employ for a similar purpose, the Hebrew legislator used NATRON, which is very common in the East. The Scriptures, in informing us that Moses made the Israelites drink this powder, show that he was perfectly acquainted with all the effect of this operation. He wished to aggravate the punishment for their disobedience, and for this purpose no means could have been more suitable ; for gold rendered potable by the process I have spoken of is of a most detestable taste. To this from GOQUET it may be well to add, that the operation of the acids which act upon gold is much assisted by the metal being previously heated. In this we see the reason why Moses cast the golden image into the fire."

Dr Memes, in his " Fine Art among the Jews," in Kitto's " Journal of Sacred Literature," vol. iii. says,—"In all the processes of metallurgy, as well as in the different departments of metallic sculpture, the artisans were practically conversant. In one of these operations, difficult even to modern science, that of calcination, Moses shows himself proficient to a degree which formerly perplexed commentators, until the more recent experiments in chemistry showed that the golden calf might have readily been reduced to a calx by burning it with *natron*, of which abundance could be obtained in the desert."

This evidently refers to the same process as the above, only Dr Memes is a little more definite in saying that the metal was first calcined. Calcination means subjecting a metal or other substance to a red heat in contact with air to oxidise it, which, notwithstanding Dr Memes' assertion to the contrary, is one of the most simple and easily performed processes in the art of metallurgy. To form a calx is to fuse the oxide of a metal either alone or with a flux, also a simple and easily performed operation. Gold, however, subjected to the process of calcination will not oxidise, and fusing it with natron has no effect upon it. The same author further says in the article above quoted from, and which is quoted in Kitto's "Daily Bible Illustrations, 1872," in favour of a theory that the calf was made of wood and covered over with thin gold :

"The facility with which such a work could be executed, suits the exigency in question, while the beauty and utility of similar artistic operations are abundantly proved by the earlier work of the Greeks. Of the Archaic specimens of this art we still possess such information as seems clearly to demonstrate that to this species of art belonged the sculpture of Aaron. Pausanias described a statue of Jupiter by Learchus, the most ancient then known, having been executed in the eighth century before our era—formed of plates of brass hammered round and fastened by rivets with a 'case' or 'foundation' of wood, exactly as the calf in the wilderness is supposed to have been constructed. Thus the earliest classic records lead us up to Egyptian practice, for to Egypt all concede the parentage of ancient art—and there we easily obtain the most probable idea of the true nature of Aaron's performance, Israel's molten calf." How a wooden calf gilt can be called a "molten calf" we leave Dr Memes to answer. Aaron says, "I said unto them, whosoever hath

any gold, let them break it off; then I cast it into
the fire, and there came out this calf." Could he
say this of a wooden figure? Common sense
forbids this; and the Doctor does not say that the
translation is wrong. But to return to the grinding of the calf.

In favour of this natron theory, Dr Eadie
speaks in a still more positive manner: "To
have drunk of water so filled with particles
of gold dust that they formed a kind of sedi-
ment in it, was no great hardship; nor can we
well understand how the ashes of the calf could
be mingled with so much water as would suffice
to be a powerful draught to the whole of the
idolaters; but if we suppose that he who was
learned in all the wisdom of the Egyptians,
employed some chemical preparation, such as was
known to the ancient world, and dissolved the
gold by means of *natron*, or other similar substance, the penalty was especially nauseous: the
smell and taste of gold so dissolved are fearfully
revolting." These last observations, as to the
dissolving of gold by natron—to the smell and
taste given by the compound formed—are so
stated that no one could doubt that the
Doctor is speaking of matters he practically
knows. At the same time, any one but slightly
acquainted with chemistry, and the properties
of gold, the whole of these references would
only produce a smile, were it not that such

F

influential writers making dogmatical statements
of this kind do great harm to Biblical truth.
Lest the natron, which is stated to be capable of
dissolving gold, might be something different
from what we had considered it, we turned it up
in Dr Eadie's "Biblical Cyclopædia," the book from
which we took the above extract, and it is there
said to be " *carbonate of soda.*"   Now, if Moses
used this substance to make the gold potable, it is
a process entirely lost.   Nothing like it is known
in the present day, and if the Doctor could define
the process, it would be a most valuable discovery.
The same chemical blunder seems to have been
made by GOQUET, quoted by Dr Kitto, but he refers to the substance used being acid, and that the
heating of the gold in the fire was to assist the
action of this acid upon the metal.   Natron is not
an acid, but the opposite—an alkali.   In order to
find out the source of these opinions that have
been so eagerly adopted, and what experiment was
referred to in these extracts that have so taken the
fancy of our Biblical commentators, after considerable research we found in Thomson's " History
of Chemistry," that Stahll, who lived in the seventeenth century, discovered that if 1 part gold, 3
parts potash, and 3 parts sulphur are heated
together, a compound is formed which is partially
soluble in water.   If this be the discovery referred
to, which I think very probable, it has certainly
been made the most of by Biblical critics.   If we are

allowed to mix up with the calf ten times the bulk
of other 'matters in order to make a grindable
or a potable compound, there are several other and
much more simple ways of doing so than with
potash and sulphur. But, unfortunately for these
explanations, they do not agree with what Moses
says he did. The account given in the Bible
will not warrant us in supposing that in the fire
there was added ten times the bulk of the calf of
other matters; and even although that had been
done, it would not have taken up one-hundredth
part of the calf, from practical difficulties in
this process. In reviewing this matter, it may
be as well first to consider what is said about the
operation in the Bible, as it should be chiefly
consulted: "And he took the calf which they
had made, and burned it in the fire, and ground
it to powder, and strewed it upon the water, and
made the children of Israel drink of it."

This last clause, "made the children of Israel
drink of it," has added to the difficulty. Metallic
gold is not soluble in water, the strewing of gold
upon the river would not affect the taste of the
water, consequently, if the water was affected by
the operation, then the gold must have been converted into a compound soluble in water. So far,
then, this would necessitate some such process as
Stahll's, not only to have it capable of being
ground, but soluble; but we do not think that
this clause, "made them drink of it," *necessarily*

means that the taste of the waters was affected by
the addition of the gold dust, but simply was an
act of contempt to exhibit the utter worthlessness
of such gods. Then a chemical compound need
not be sought for to explain the difficulty. The
mere act of grinding, either in a mortar, or
between stones, could be easily effected by the
addition of a very minute quantity of lead, or tin,
to the gold, which renders it quite brittle and
capable of being ground to powder; but we do
not think that even this was done. In the recapi-
tulatory book a more definite account is given by
Moses of the operations by which this grinding
was effected:—"And I took your sin, the calf
which you had made, and burned it with fire, and
stamped it, and ground it very small, even until
it was as small as dust, and I cast the dust
thereof into the brook that descended out of the
mount " (Deut. ix. 21).

In this account there is no indication of
solubility, but rather the opposite; because if it
were soluble, and intended to make and keep the
waters bitter for a time, there was no necessity
for grinding it so fine as mentioned; but apart
from any supposition, let us follow this plain
statement given by Moses himself of the operation—

1. It was put into the fire, no doubt, so that it
might be cast into bars or ingots of suitable
size for the operations to follow.

2. "*And stamped it;*" that is, beat it out into thin laminæ or leaves. It is well known how thin it is possible to beat out gold—much thinner than the finest paper now made; and Wilkinson, in his "Ancient Egypt," says that the Egyptians were well skilled in the art of gold-beating, consequently so would Moses be, as well as many of the Israelites, as the following passage proves: "And they did beat the gold into thin plates, and cut it into wires (or threads), to work it in the blue, and in the purple, and in the scarlet, and in the fine linen and skilful work."

3. "*And ground it very small, even until it was as dust.*" Gold-leaf placed between stones or in a mortar, and ground, can be thus reduced to fine powder, fine as dust—a little oil, honey, or any unctuous liquid moistening it, facilitates this operation. Extensive manufactories exist in England performing these same operations under a patent at the present day, converting malleable metals and alloys into fine powders, fine as dust, known in commerce as metallic bronze; and so fine is this bronze, that it resembles more the dust from the wings of a moth or butterfly, than metal. The laminar structure of these dust particles will cause them to *float in water for hours, and in a running stream for days;* so that even Dr Eadie's difficulty of not knowing "how the ashes of the calf could be mingled with so much water as would suffice to be a painful

draught to the whole of the idolaters," is thus easily got over—even allowing that it was intended for the people to drink the dust. Considering that these operations were performed by hand in the wilderness, it must have been very humiliating for the idolaters to continue from day to day, probably from week to week, as it would require much time, in the presence of the people, *pounding, stamping*, and *grinding* their god before which they had so recently danced, played, and bowed themselves, as the God who had brought them out of Egypt. Contempt could not be more complete.

There is a singular idea handed down by tradition, that the beards of all those who drank of these waters took the colour of gold. Bochart refers to the fragment of a version of Exodus begun in the thirteenth century, in which the 27th verse of chapter xxxii. is thus translated, " Slay every man his brother, his friend, his neighbour, namely those who have the golden beards." An explanation or gloss is added, " Those who worshipped the calf had their beards gilded ; for the powder was stuck there by miracle." Whatever was the origin of this tradition, it shows that in early times they had no such idea as our modern commentators have of Moses converting the gold into a salt for solution.

The grinding of idols and altars that had been

used in idolatrous worship, was repeatedly practised afterwards, as is stated in several parts of
Scripture. Rawlinson, referring to this practice,
says, in reference to Josiah burning and stamping
the idols, and casting the powder on the graves of
the children of Israel, that these were no doubt
metal, and agrees exactly with what Moses did
with the calf. Josephus, in the 7th chap. of Book
viii., states that the fashion of powdering the hair
was practised in Solomon's days, and that the
riders of his horses were accustomed to powder
their hair with gold dust. In all probability
this powder was obtained as we get our bronze
powders, and as Moses ground the calf.

After publishing our little work, "Ancient
Workers in Metal," our remarks upon the grinding of the calf were found fault with by Dr Eadie,
who took the opportunity in a foot-note in his
"Life of Dr Kitto," to give us a severe rebuke for
our audacity in venturing to question the truth of
his statements, and after a sneer at our inability
to explain how Moses ground the calf, leaves his
readers with the impression that the grinding was
done by some wonderful unknown process. We
replied in a letter to the Editor of the "Journal of
Sacred Literature," and, as it is calculated to bring
out the matter more fully, we give the letter,
and leave the verdict to the decision of our
readers.

" Sir,—In a recent number of your journal, my little book upon Ancient Metallurgy was favourably noticed. In that work I endeavoured to show that references made to metallurgical operations in Scripture, generally agree with the operations for effecting the same purpose in the present day. Connected with these metallurgical subjects is the demolition of the golden calf by Moses, which has long been termed the commentators' puzzle.

"Amongst several others who had ventured, or rather adopted, conjectures upon this puzzle, I made particular reference to the Rev. Dr Eadie, who—apparently adopting an old conjecture that Moses fused the calf with sulphur and an alkali, and produced a grindable and potable compound —has represented Moses as mixing and fusing it with carbonate of soda, and considers the process a familiar operation.

" Instead of following these impractical and absurd conjectures, or searching to find some process that could be applied without reference to the text, I considered Moses' plain and graphic statement, and showed that it exactly corresponds with what we, in our manufacturing operations, really do at the present day for effecting the same object—grinding metals to powder—and contented myself with this, as being at all events a consistent and practical process. However, Dr Eadie still thinks differently, and in his recent ' Life of Dr Kitto,' I have been honoured with a special

notice in reference to this matter, which I will
now quote :—

"Another recent author has taken up and rebuked both
Dr Kitto and ourselves upon a point on which he possesses
practical skill and experience. The matter in dispute is
the demolition of the golden calf by Moses. The conjecture may be untenable that Moses dissolved the calf in
some chemical fluid, and mixed the nauseous potion with
the water which he compelled the Israelites to drink,
though certainly a solvent sufficient for the purpose might
easily be fixed upon, and might be known to Egyptian
chemistry. The words of Moses are, 'He burned it in
the fire, and ground it to powder, and strewed it upon
the water, and stamped it, and ground it very small,
even until it was as small as dust.' Mr Napier thus
explains the process,—'It was put into the fire that it
might be cast into bars suitable for the operations which
were to follow.' Thus the modern chemist says it was
melted, but Moses declares it was 'burned' in the fire ;
not melted, certainly, for the different language plainly
describes a different process. By stamping, Mr Napier
understands beating it out into thin leaves, a refinement
of operation which the words do not warrant. The text
implies that the burning was not fusion, but some unknown process that prepared the metal for the stamping
and grinding, a process which Mr Napier, though he
meditated a book on the Chemistry of the Bible, has not
discovered, but has been obliged to leave unexplained.

" That Dr Eadie holds a high position in his own
sphere of labour, and is an eminent authority in
that field, no one can doubt, but on matters of
practical science it is evident he is not sufficiently
informed. The Doctor says that 'the conjecture

may be untenable that Moses dissolved the calf in some chemical fluid, though certainly a solvent sufficient for the purpose might easily be fixed upon, and might be known to Egyptian chemists.' This is true; the dissolving the calf in a chemical fluid is untenable, not for want of one being discovered, as solvents sufficient for the purpose are known, and we are now all but certain were also known to Egyptian chemistry, but the text will not admit of a dissolving process, however practical and well known. The processes referred to in the text are mechanical, not chemical.

" Metallic gold cannot be combined with another solid body in the fire without one of them being melted, and the whole of the compound formed by their combination being in a state of fusion afterwards, so that Dr Eadie's critical examination of the original strikes at the root, not only of his own, but all the conjectures of the same kind which have been made. It is to be regretted that the Doctor did not define what is really meant by the word ' burn,' which he so positively says cannot be applied to fusion. The operation of burning may be defined as the combination of a body with oxygen when subjected to the heat or fire. Thus a piece of coal or wood put into a fire burns, and the compound formed between the oxygen and the body burned being a gas, the whole is dispelled. Under these circumstances, burning is identical with complete destruction.

Certain metals, such as iron and copper, put into a fire, also burn, but the compound formed between the metal and oxygen being solid, is a crust upon the surface of the metal which protects from further burning. By removing this crust, burning goes on to form another crust, and so on until the whole of such metals may be burned through time, and these metals are by this process as fully destroyed as the coal or wood. Had the calf been made of such burnable metals, the word burn, taken in this sense, would have been applicable, as being within practical possibility, although the text, as translated, hardly allows this mode of destruction. When this definition of the word burn is applied to gold it is untenable. It is one of the distinguishing properties of gold to resist the fire, which means being incapable of burning. Indeed, when gold is combined with oxygen, and this oxide put into a fire, the metal is revived, and the oxygen given off, and fire has ever been the practical means for purifying gold, as is evident from Scripture. So prominent and well known is this law, that to burn gold by putting it in the fire would be as great a miracle as causing the head of an iron axe to swim above water; and we have no idea from the text, that Moses in burning the calf wrought a miracle, or wished the Israelites to believe so, but rather that he performed a well known operation. May not the word *burn* have a more popular sense than

that here given, and evidently meant by the
Doctor's remarks, as it appears to have in other
passages of Scripture, *i.e.*, Deut. iv. 11, ' And
the mountain burned with fire ? ' The term
is popularly applied to making a body red hot.
However, the true rendering of the original, and
the meaning evidently applied to the term burn,
I leave to scholars as an important inquiry.

"The objection which the Doctor makes to
hammering the gold into thin leaves previous to
grinding, as being too refined, or rather that
the language will not warrant so refined an
operation, I only mention, because many of our
most refined operations in metals were familiar to
the ancient Egyptians, and consequently would be
referred to incidentally in the most popular
manner, as seems to be done in the text under
review. As to the ability of the Egyptians to
beat gold there can be no dispute.

" As to the Doctor's remarks upon my meditating
a book upon the Chemistry of the Bible, while I
had not discovered how Moses burned the calf, I
would merely say that the demolition of the calf
did not come within the sphere of chemistry, and
I believe no chemist has been capable of dragging
it into that false position. I understood, and still
believe, that the process was wholly mechanical ;
and it is astonishing how men professing a belief
in the record, and yet wise beyond what is written,
lay aside a plain description of a process done

under the eyes, and by the direction of the writer of it, and say, 'No, that is not what you did; you must have dissolved it in some chemical fluid you were told of in Egypt, or melted it with some such substances as sulphur and an alkali, or, perhaps, *natron* being plentiful in your country, you have melted it in this; you were not mechanical enough or sufficiently refined in your operations to do it in the way you have stated.' Such is really the language that has been used in reference to this passage.

"I would say, in conclusion, that I do not advocate any particular mode of demolishing the calf. All I wish to show is the identity of the process described by Moses, as translated in the authorised version, with the practice of the present day, and that if we were called upon to perform the same task, we would adopt the same method as Moses describes. I would add, that if commentators, however high they stand in theological science, would, when they fearlessly, often recklessly, venture into the field of practical science, adhere more closely to the text, there would be fewer puzzles; or probably, what would be a wiser course, if they would let all matters requiring the knowledge of practical science alone, if they have not personally studied them, Scripture truth would certainly gain and not lose.—I am yours, &c.,   JAMES NAPIER."

COPPER.

THE next metal the metallurgy of which we are now to consider is copper. The name copper (*cuprum*) is said to be derived from Cyprus, an island in the Mediterranean originally peopled by the Phœnicians, because it was a rich mineral district. Pliny and Strabo mention that in their time this island was celebrated for minerals, among which are mentioned a variety of metallic ores.

Copper is found in great abundance in nature in a mineralised state, that is, in combination with other substances, the more common of which are oxygen, sulphur, and carbonic acid; these mineral combinations are termed ores, and the copper is separated from the other substances by smelting. The various copper ores have each distinct characteristics; they are either very heavy, or beautifully coloured purple, blue, or green, and would on account of these colours attract very early attention. It is occasionally found in nature in a metallic state so pure as to be used at once for manufacturing purposes, either for making articles of copper or alloys. There are abundant instances of this in modern times where masses of metallic copper are found. At the mines at Lake Superior, in America, a mass of copper was found a great many tons in weight; it is therefore probable that quantities of native copper were also found in mines in ancient times, so that the ancients could possess this metal without the necessity of smelting it from ores. But judging from modern times the quantity thus found is upon the whole so trifling, it would scarcely be worth naming as a trade, and if even such huge masses were found as at Lake Superior, it would certainly be a work of great difficulty to reduce them to a size suitable for the furnace or smelting-pot, as they cannot be broken like stone, but have to be cut, requiring tools of particular hardness, and other appliances and operations more difficult than the smelting and obtaining the metal from the ore. The copper so obtained must have formed but a very small portion of the supply, as we may infer from the extensive use made of that metal both in the arts and for common articles of use. It must, therefore, have been procured from the ore, and that of necessity implies a knowledge of smelting. We read in Job, "Copper is molten out of the stone," a direct intimation that the ancients were acquainted with that process.

Before proceeding to describe the smelting process, we will notice what references are made to copper in Scripture. In our translation the word only occurs once, and that is when enumerating the various articles brought by the Jews from Babylon into their own country. Ezra says, " Also twenty basins of gold of a thousand drams, and two vessels of fine copper, precious as gold." Almost all Biblical critics agree that these vessels were not made of copper, but of a rich alloy capable of taking on a bright polish. This we think highly probable, as copper was then in common use amongst the Babylonians, and would not therefore be "precious as gold." Some commentators, without due consideration, however, have supposed it probable that those vessels were made of *Corinthian copper*, or *Corinthian brass*, an alloy of peculiar richness, found after the burning of Corinth among the ashes of some of the temples, and formed by the fusing of various metals together. But as the Jews returned from Babylon in the year 536 B.C., and Corinth was not destroyed till 146 B.C., nearly four centuries after, it is impossible that these vessels could have been made of that alloy. Others with greater probability think they were composed of an alloy of gold and other metals much esteemed amongst the Persians, which took on a high polish and was not subject to tarnish.

In reference to the copper of Scripture, it must be remembered that the translators, not being acquainted with the technicalities of metallurgic art, did not understand the distinctions which names bear in the arts; hence we find the word brass used synonymously with copper and bronze.

Brass is a compound of copper and zinc; bronze
a compound of copper and tin—alloys of distinct
character and composition, and of artificial production, there being no such things as a *brass* or
bronze ore. Consequently, when we read, "Brass
is molten out of the stone," or, "Out of whose hills
thou mayest dig brass," it is evident that it is not
the alloy brass which is meant, but the metal or
ore of copper. Besides, there is no evidence in
Scripture or in other writings, nor has any remnant
of ancient art been found to prove, that the metal
zinc was known at all to the ancients. In most
other instances the word brass should be translated
bronze; and as copper is the principal metal in
this alloy, it follows that a reference to bronze
necessitates a previous knowledge of the metallurgical operations for copper. Bronze was an
alloy well known in the earliest times. Indeed,
among the antediluvians we find mention of
Tubal-cain, "an instructor of every artificer in
brass and iron." From this, though it is not mentioned, we may infer that the knowledge of metallurgy, and the art of working such metals as iron
and copper, greatly facilitated Noah's operations,
both in the preparation of materials for the ark,
and the construction of that monster vessel itself.

We will now describe the methods of extracting
the metallic copper from the ores by smelting,
and point out practical difficulties in the process.
We mentioned in a preceding part that gold and

G

silver, when in union with oxygen, gave off that oxygen when heated to redness; but if oxide of copper be brought to a state of redness, the same effect is not produced; on the contrary, the oxygen is more firmly fixed. So strong is the attraction of copper for oxygen at this temperature, that if pure copper be heated to redness, it combines with oxygen, forming a black crust. If, however, when the oxide of copper is at a high temperature, carbonaceous matter, such as charcoal, be brought into contact with it, the carbon combines with the oxygen, forming carbonic acid gas, which flies off, leaving the copper in the metallic state. Hence, when an ore-compound of copper and oxygen is mixed with coal, and brought to a bright red heat in a furnace, the copper is reduced to the metallic state, and falls to the bottom, where it can be run off by an opening made for the purpose at the bottom part of the furnace.

So that, so far as this kind of ore, namely, the oxide, is concerned, the process is so simple that even the rudest people living in the neighbourhood of such ore could not fail to know the means of extracting the metal from it. Should the metal be combined with carbonic acid, forming the extensive class of ores termed carbonates, such as malachite, azure copper, &c., the same operations are applicable to them as described. The carbonic acid flies off at a dull red heat, and leaves the copper in a state of oxide; so that by the time the heat has risen sufficiently to melt the copper, the ore is in the condition of oxide, and the copper will become reduced and melt, as just stated. These two classes are considered to have been the principal ores smelted until a comparatively late date; nevertheless, we do not see why we should limit the ancients in matters, the knowledge of which is so closely allied to what they did know, and which we find so simple.

When the copper is combined with sulphur, forming what is termed a sulphide ore, it is first subjected for a considerable time to a dull red heat, which drives off the sulphur, oxygen taking its place, and converting the copper into an oxide. This operation of burning off the sulphur is done in different ways, first breaking up the ore into small pieces, and piling it up into a large heap, with air openings underneath, the ore being mixed with coal, turf, dried dung, or other carbonaceous matters, to maintain the temperature. These heaps generally amount to several hundred tons. When the heap is first made, it is covered over with clay or turf; a fire is then kindled at one end, which, communicating with the carbonaceous matter throughout the heap, induces a slow combustion of the mass, which continues burning for a period of from ten to twelve months, during which the sulphur is not only driven off, but all the metals present are oxidised, and the earthy matters also brought to a condition of greater fusibility. This operation

is termed calcining; and so destructive to plant-
life are the fumes given off, that no vegetation
can exist for a long distance around these heaps.
Recently this operation has been performed in
large reverberating furnaces constructed for the
purpose.

The method of calcining in heaps in the open
air is one that has been employed from very
early times, and is still practised in the Island of
Anglesea, and on different parts of the continent.
When a heap of ore is calcined, and the sulphur
well burned off, the continental practice is to trans-
fer the ore to a cupola furnace, and mix it with
fuel, and when it is fused, an impure copper is
obtained. If a portion of sulphur still remains in
the ore, a quantity of copper will be found combined
with it after fusion. This is subjected to another
burning, until the copper is free of sulphur, and
is again fused; the impure copper is afterwards
purified by refining. In this country, where cupola
furnaces are not used, the calcined ore is put into a
reverberatory furnace, and fused; but as the sulphur
is seldom, perhaps never, entirely taken out of the
ore by calcining, and as fusion merely serves to separate the earthy matters and other impurities, the
copper and remaining sulphur form a compound
by themselves, which has afterwards to be subjected to fire and a current of air to remove the
remaining sulphur, after which the copper is
obtained by fusion. If all the sulphur be driven

away before fusion, then carbonaceous matters are mixed with the oxide of copper remaining, and copper is obtained as described for the oxide; but in the present day, in this country, the sulphur is purposely not all driven off, the copper being obtained without the use of carbonaceous matters.

The more common method at present in use for obtaining the copper from sulphide ores is as follows :—The ores always contain a large quantity of earthy matters, the copper present being generally little more than 10 per cent of the whole stone. The ore is broken into small pieces, and submitted to a low red heat for several hours in a large reverberatory furnace, with a current of air passing over the surface of the ore, which is turned over every now and again, to expose a new surface to the current. This process burns off a large portion of the sulphur, and oxidises some of the iron which is always present in such ores. This calcined ore is then transferred to another reverberatory furnace of lesser size, and fused, during which process the earths and oxidised iron combine, and form scoria or slag. The copper then combines with the remaining sulphur and iron, and the mixture, by reason of its greater gravity, sinks under the scoria. The slag is then skimmed off with a rake, and the metallic portion run from the furnace into a pit of water, and thus granulated. This granulated mass is again calcined, and fused as before; and so on, alter-

nately roasting and fusing, until the iron and
most of the sulphur are burned off. The copper
and sulphur left are then existing in the condition
known to chemists as subsulphide. This is then
submitted to another roasting at a higher heat,
when, by a beautiful chemical reaction between
the sulphur and oxygen of the air, the copper is
obtained in the metallic state.

This beautiful, although somewhat tedious, process, the result of long experience, has now
become a necessary process for working such ores
of copper as are found in this country, having in
them different metals, which by their oxidation
and fusing with carbonaceous matters, would
combine and deteriorate the copper; but by the
process described these metals are scoriated, and
the copper and sulphur left pure, to be treated
as above. These operations require much practical
skill on the part of the workmen employed, and
about eight days' constant work before the metal
is obtained pure and fit for market.

Whatever may have been the process adopted
by the ancients for the extraction of copper from
the ore, there are certain practical difficulties,
arising out of the nature of the metal as well
as the character of the ores, which must have
existed in ancient times as well as at the present
day. These difficulties must be overcome before
the metal is in good condition for being manufactured into objects of art. Thus copper ores,

as we have already stated, especially the sulphide ores, have small portions of other metals mixed with them, which are capable of being reduced by carbon; and these being fusible at a temperature of melted copper, consequently become alloyed with the copper, and render it impure. When this is the case, the copper must be kept at a melting-heat, and exposed to the air; and the metals alloyed with it, oxidising more rapidly than the copper, are consequently burned or roasted off. Much copper, however, is lost by this method. Another process may also be adopted : the alloyed copper may be fused, and a quantity of pure oxide of copper added; this will tend to oxidise the deteriorating metals, and purify the copper. The first of these processes is the one adopted at the present time, and the copper, burned or oxidised off, is recovered by other operations.

Again, pure copper, when melted, absorbs oxygen and other gases, which render it hard and brittle; and although it may be used for casting, it cannot be used for hammered or rolled work. Now, the method of overcoming this difficulty is by plunging into the melted copper a piece of green wood, and holding it in this condition until a portion of metal taken out, and tested by hammering and breaking, shows it to be in a malleable state. This operation is termed refining, or toughening.

Mr E. H. Palmer, in his " Desert of the Exodus,"

speaking of the mineral wealth, says—" This
formation is rich in mineral wealth, containing
many veins of iron, copper, and turquoise. The
absence of all convenience for smelting and
transport deprives them of commercial value at
the present day; but the ancient Egyptians appear
to have had greater facilities, and to have worked
the ores upon an extensive scale. The neighbourhood of Seráabit el Khádim and Maghárah abounds
in mines, in hieroglyphic tablets recording the
names and titles of the kings under whose
auspices they were wòrked. One of the tablets
found mentions the " Goddess of copper " as
being the presiding deity of the place. From the
inscription and cartouches found there, it is
evident that the mines were in full working order
at the time of the Exodus, and that the neighbourhood of Seráabit el Khádim must have been
occupied by a large colony of workmen, and
probably a considerable military establishment to
preserve discipline, as the miners were chiefly
selected from criminals and prisoners of war.

  " There are abundant vestiges of large colonies
of Egyptian miners, whose slag-heaps and smelting furnaces are yet to be seen in many parts
of the peninsula. These must have destroyed
many miles of forests in order to procure the fuel
necessary for carrying on their operations.

  " The instruments employed I believe to have
been chisels of bronze, or other hard metal,

and not the flint flakes which are found in such quantities in the vicinity. The Egyptians, we know, were expert metallurgists, and flint implements could hardly have made such marks as those on the stone. The country around the mines contains numerous evidences of the immense smelting operations carried on by the ancient Egyptians."

Speaking of another district, called Ragasta, and its copper mines, he says—"A large dyke runs through the granite along the top of a low ridge of hills, and contains thin layers of the metal in a very pure form. The grain of the rock itself also contains a considerable quantity of the ore in minute particles; but the miners seem to have been ignorant of any method for crushing the stone, and seem to have contented themselves in picking out the thin layers of sulphate (sulphide?) of copper from the dyke. At the end of the ridge the ore has been worked out in a small cave; and in one place where the vein takes a dip, a shaft has been sunk to a considerable depth. The neighbouring hills are covered with pathways in every direction; and the numerous remains of smelting furnaces which may yet be seen show that mining operations were once carried on upon a very large scale in the vicinity."

These observations would have had their value greatly enhanced had Mr Palmer given a description or sketch of the remains of the furnaces which were used in smelting, and also obtained analyses of the slags and ores. To show of how ancient date is the mining and smelting of copper ores, we extract the following notes from an article by H. Bauerman, on the "Geology of Arabia Petria," read to the Geological Society of London in 1868 :—

"At Wady Khalig, a tributary of Wady Baba, about four miles below Nasb, the iron and manganese bed has been extensively excavated by the old miners. The old workings extend about 120 yards along the face of the hill; the walls and pillars are covered with small chisel or gad marks, apparently made with a tool about three-fourths or seven-eighths of an inch in breadth. There are no inscriptions, or any other guide to the probable date of these workings; but it is evident, from the extraordinarily poor character of the ore, that they must belong to a very early period, when metals were of uniform value, owing to their production being confined to a few localities.

"The ores from the mine at Wady Khalig were smelted in Wady Nasb, close to the springs, as is evidenced by the mass of slag which forms a roughly elliptical heap about 350 yards in length and 200 in breadth. The depth is very variable, and probably not more than eight or ten feet at the most; and over the greater part of the area the slag forms only a thin covering to the rock. Upon digging into the heaps, numerous clay twyer-nozzles, with an air-passage of about three-quarters inch diameter, were found. These have evidently been formed by the accretion of the slag to the wall of the furnace in front of the blast, and in many cases have been repaired by plastering fresh clay over the former face—an operation which is seen from some of the broken twyers to have been repeated three or four times. In one instance the slag nose was accumulated to such an extent as almost completely to block up the passage for the blast. In nearly all cases, shots of metallic copper are found included in the slag adhering to the twyers, but not a fragment of unaltered ore, or any kind of regulus, was found, although carefully looked for. The only building that can be regarded as having formed part of a furnace is a pair of small walled enclosures of unequal size, the larger one being about six feet square, and the other two and a half feet square, both being walled in on three sides to a height of about two feet from the ground. These may possibly have been the outer walls of a hearth or low-blast furnace, but no trace of any lining that had been subject to the action of heat could be detected. In the smaller compartment a stone pestle, worked round, and about one and a half inches in diameter, was found.

" On examining the different parts of the heap, it became evident, from the nature of the slags themselves, that several different operations were carried on here. Thus in places the fragments are broken up small, and contain many shots of metal, now mostly changed to malachite. These are probably rich selected cinders, either from the first fusion, or perhaps from the refinery, which have been put on one side for further treatment; while on the other hand, at the upper end of the heap, crusts of well-smelted, clean slag, from one and a half to two inches in thickness, are spread over the ground, as though they had been allowed to flow from the furnace, and solidify on the rock in the place where they are now found. It may well be, therefore, that these represent the operations of larger furnaces, worked perhaps at a later period, when the art of metallurgy was further advanced than was the case when the thinner and less perfectly melted slags were produced.

"At the lower end of the Nasb valley there is another slag-heap of smaller extent, but in other respects similar to that of the springs. A third was found in Wady Gharandel, upon a terrace of limestone, far away from any place producing copper ores, but near water, proving that the sites for smelting-works were determined chiefly by the presence of springs, where there is usually some quantity of wood even at the present time. The charcoal used for smelting appears to have been derived from the *acacia vera*, which still flourishes round the springs. I was unable to find any traces of the ruins of large calcining furnaces and basins that had probably served as stamps and catch-pits, as described by Dr Figari Bey."

Speaking of the ancient miners' tools found in the old workings in Wady Maghàrah, he says— "The old faces of the works bear irregular tool-marks, which leave no doubt that they have been made with flint flakes, great numbers of which are found strewing the valleys and hill-sides, and within the workings. Most of the flakes are of a triangular section brought up to a point, which is generally well worn and rounded, and the shape of which when blunted corresponds perfectly with the grooves on the face of the rock."

He also found fragments of stone hammers in the workings; and from other fragments of tools, and small cylindrical blocks of wood, having a notched head to receive a withe or cord, he thinks that these were mountings for the flint chisels employed by the ancient miners; for without something of this kind it would be difficult to work with the flakes, owing to their tendency to break across when not struck fairly on the top. The hammers found in the workings were of a very rude kind, in many cases being rough fragments of dolorite.

"The period over which the working of the mines at Wady Maghàrah extends, according to the evidence of the numerous hieroglyphic tablets

covering the face of the cliff, as interpreted by
Egyptologists, extends from the 3d to the 13th
Manethonian dynasties, corresponding to an interval of about 1600 years.   There seems little or
no difference of the work done in cutting the
tablets of the oldest and the newest, so that there
seems to have been little progress made in the art
of cutting stone during that long period; indeed,
the oldest is the best."

These extracts are very suggestive.   In the
first place, as to the date of these smelting and
mining operations, when was the third Manethonian. dynasty ?   Having consulted authorities
and calculated back, we find that the year 812
B.C. occurs in the 25th dynasty, and 1920 B.C. is
the date when Abraham is said to have visited
Egypt, which was in the 15th dynasty.   Between
these two periods there are 1108 years, equal on an
average to 110 years to each dynasty.   If we allow
110 years for the 14th dynasty, and add the
1600 years given above from the 13th to the 3d
dynasty, we have—

From 3d to 13th dynasty,	.	.	1,600
„ „ 14th „	.	.	110
From Abraham,	.	.	.	1,920
We have,	.	.	.	.	3,630 years B.C.

as the date of the first tablet when the copper
mines were wrought.

In the second place, we have here a period of

1600 years, during which these mining operations were in progress, when there were nothing but stone tools used of the rudest sort. According to a rule laid down by certain archæologists, these people lived in the stone age; and yet we see that all this time they were mining and smelting copper, which in all probability was carried away to Egypt, and used for making bronze. Is it not likely, as suggested by Mr Palmer, that these miners were banished felons and prisoners of war? and that, like other nations in ancient times, Egypt retained the prisoners who were skilled workmen to enrich their cities, and those who were not skilled sent to the mines? and that to make their lot more arduous, they were not allowed other kind of tools but those they themselves could make from stone, and yet were obliged to do their "tale" of work, as was the case at a later date, when the Israelites were refused straw, and had to gather it themselves? These researches prove the important fact, that even in these very ancient times copper, and no doubt other metals also, were smelted, and their value highly appreciated, by the Egyptians. There are no references to copper smelting in Scripture except the fact mentioned in Job that "copper was molten out of the stone," so that we cannot say what methods were practised by the Jews. But there is little doubt that the same difficulties occurred then as now; and if they did not overcome them through the same methods exactly as we do now, they must have employed others equally efficient, otherwise their copper would not stand working.

Again, in making copper alloys that are to stand working by hammer, or to be rolled into sheets, the copper used must be of the purest quality.

Illustrations of this principle are numerous. One may be cited by referring to the well-known alloy termed *yellow metal*, which is so extensively used for sheathing, boiler pipes, and engine work. The copper for this alloy is made expressly for it, and termed *best select;* and if anything be wrong with the copper, either through imperfect refining or through having an alloy in it, it will not suit. The same is the case with bronze guns, and such articles of that alloy which have to stand great wear and fatigue. The copper must be of the best quality for success.

Many of the ancient copper alloys had to stand working by the hammer, and their bronze was such, in respect to toughness or hardness, that we cannot at the present day make anything to surpass it. This is surely strong presumptive evidence that the copper, as well as the tin they used for these alloys, was pure, and that they understood the methods of effecting this object. This is further proved by some analyses made of ancient bronze, which will be found in another part of this work.

In searching for evidence of the modes of smelting copper, or any other metal in ancient times,

there is very little data to be found upon which
we can rely. Such arts were confined to guilds,
and were kept secret, the hard manual labour being
performed by slaves, prisoners of war, and criminals,
so that written descriptions are not to be found,
and the writers of history too often despised such
arts, and consequently they did not notice them
with any minuteness; hence, as already stated, we
are much more indebted to the figures and alle-
gories of the poet, explained inductively, than to
any direct description. In our opinion, the first
method of smelting ores was by heaping together
wood and ore, and burning them in masses. In
Macedonia, where, in the time of Philip, the father
of Alexander, lead mines were worked and the
ore smelted, large heaps of slag are still found,
so far above the level of the rivers of the country,
that the furnaces in which they were produced
must have been blown by bellows, or probably so
arranged, like the old bloomeries, that they were
blown by the wind.

Pliny mentions that King Numa, the immediate
successor of Romulus, founded a fraternity of
brassfounders (bronze-workers). This statement,
if correct, shows that copper, which is the principal
constituent of bronze, was plentiful. The same
writer states that two distinct kinds of copper
were exported from Cyprus, one called *coronarium*,
which, when reduced into thin leaves and coloured
with oxgall, had a golden colour, and was em-

H

ployed for making coronets and tinsel ornaments
for actors, from which circumstance it derived its
appellation.   Another variety, which was named
regulaire, is not particularly described; like the
former, it could stand hammering.

It is also mentioned that in France it was usual
to melt copper among red-hot stones, for the purpose of obtaining a steady heat, as a quick fire
was found to blacken the metal, and render it
brittle.   Pliny states that the process was completed in one operation, and that the quality would
be improved by more frequent melting; he also
remarks that all kinds of brass melt best in cold
weather.

Aristotle tells us that the Mosynæci, a people
who inhabited a country not far from the Euxine
Sea, were said to make their copper of a splendid
white colour, not by the addition of tin, but by
mixing it with an earth found in the country.   It is
probable that this earth was calamine or zinc ore,
and that the metal obtained was really brass.   But
these remarks give us no idea of how the smelting
operations were performed ; and when Pliny and
other ancient writers do refer to practical arts,
they are not very correct in their description of the
processes.   And the translators of their works have
added to the confusion, either through ignorance,
or on account of the poverty of the original language in technicalities, as we find brass in one
place, copper in another, white copper in a third,

all used indiscriminately, without regard to whether the original referred to pure copper, or to copper whitened by the addition of tin or lead, or by any other process. This being the case, we must use inductive reasoning, and more especially as a great part of our present inquiry refers to a much earlier date than the days of Pliny.

It does not appear that pure copper alone was very extensively used by the ancients, either for ornamental purposes or for use. This may be accounted for by the difficulty of toughening pure copper so as to fit it for the hammer; yet the fact that copper vessels made by hammering have been discovered amongst ancient relics, shows that they were not ignorant of toughening. It has been mentioned that toughening is effected by inserting poles of green wood into the melted copper. This is for the purpose of removing any oxygen which may have combined with the metal, and which makes it brittle. But other matters may be used instead of wood. If a small portion of lead or tin be added to the copper, these will effect the same purpose, although preventing the copper from being entirely pure, as any excess of tin or lead above that required to take out the oxygen will remain in the copper. If the quantity is small, this would not affect the quality of the metal, either for use as copper or for bronze, for which copper seems to have been principally used by the ancients. And that this mode of toughening was known to

the ancients is more than probable from their
great skill in manufacturing bronze. The best
way to determine this question is by making
analyses of ancient copper, for, as a general rule,
copper so toughened or deoxidised will contain
small portions of tin or lead, whichever metal has
been used. The following are the only analyses
we can find:—

Statues of copper horses erected in ancient
times; date not known. Analysed by Klaproth—

Copper,	.	.	.	.	.	99·3
Tin,	.	.	.	.	.	·7
						100·0

Broken spear-head. Analysed by J. A. Phillips.
No date—

Copper,	.	.	.	.	99·71
Sulphur,	.	.	.	.	·28
					99·99

Coin, A.D. 262. Analysed by J. A. Phillips—

Copper,	.	.	.	.	97·13
Tin,	.	.	.	.	·10
Silver,	.	.	.	.	1·76
Iron,	.	.	.	.	1·01
					100·00

Coin, A.D. 267. Analysed by J. A. Phillips—

	No. 1.	No. 2.
Copper,	98·50	98·00
Tin,	·37	·51
Iron,	·46	·05
Silver,	·76	1·15
	100·09	99·71

These analyses are too few to warrant us in drawing conclusions as to the mode of refining and deoxidising. Besides, they are mostly samples of cast copper, where the toughening process was not necessary. One reason why copper was not extensively used in a pure state for utensils was on account of its tarnishing quickly, thereby producing a poisonous matter upon its surface, unfitting it for domestic purposes. Michaeli makes the observation that Moses seems to have given a preference to copper vessels over earthen, and on these grounds endeavours to remove the common prejudice against their use for culinary purposes. But this is a dangerous conclusion, and no doubt arose from that confusion of terms referred to of calling bronze copper. We are inclined to think that Moses used no copper vessels for domestic purposes, but bronze—the use of which is less objectionable, it not being so apt to tarnish, and takes on a finer polish, and also being much more easily melted and cast, would be more extensively used than copper alone. Nevertheless, there is evidence in the remains of articles that have been found in old ruins, dating probably before Moses' time, that copper was used for purposes to which we cannot apply it, showing that the ancients possessed a certain knowledge of how to use copper tools which we have lost.

In describing one of his excavations in ancient Chaldea, Mr Loftus says—" Following the wall for

a distance of six feet, the workmen discovered a
number of copper articles arranged along it, which
form a very curious and quite a unique collection,
consisting of large caldrons, vases, small dishes,
and dice-box(?), hammers, chisels, adzes, and
hatchets; a large assortment of knives and daggers
of various sizes and shapes, all unfinished; massive
and smaller rings, a pair of prisoner's fetters, three
links of a small chain, a ring weight, several
plates resembling horse-shoes, divided at the heel
for the insertion of a handle, and having two holes
in each for pins; other plates of a different
shape, which were probably primitive hatchets;
an ingot of copper, and a great weight of dross
from the same smelted metal. There was likewise
a small fragment of a bitumen bowl overlaid with
thin copper, and a piece of lead. The conclusion
arrived at from an inspection of these implements
and articles is, that they were the stock-in-trade
of a coppersmith, whose forge was near at hand;
but the explanation of their connection with the
public edifice near which they were found is by no
means clear. They are well and skilfully wrought.
One of the hatchets particularly attracted my
attention, being of the same form as that represented on the tablet of the man attacking a lion.
The articles which I conceived to be dice-boxes
precisely resemble those of modern form. The
knives were all adhering together *en masse*, their
rough, broad edges proving that they were never

finished by the cutler. The total absence of iron in the older mines implies that the inhabitants were unacquainted with that metal, or at any rate, that it was seldom worked." With this last remark we do not agree. The absence of metallic iron in connection with copper is not to be wondered at, for besides the rapid destruction by oxidation of iron articles lying in the earth, this destruction is hastened vastly by the presence of copper when in contact, by the generation of a galvanic action. Probably, however, iron was very scarce in Chaldea, and only used along with copper, perhaps, as Wilkinson suggests, to tip their chisels, &c.

After reading the above, we wrote to the author, and asked if any of the specimens had been analysed, and if there was any tin in combination with the copper, to which he kindly replied, and referred to specimens in Mr Vaux's room in the British Museum, and also stated that some of them were being tested at the School of Mines, and remarked—

" It will be extremely interesting and curious to ascertain that they contain tin, because those from Tel Sefr are of so much earlier date than those discovered by Layard at Nimroud. There being no tin that I am aware of in the Persian mounttains, can it be possible that the metal could have been derived from Britain at such remote date (1500 B.C.) ? "

We afterwards learned that a sample of these curious relics was being analysed by Mr Spiller of the Royal Arsenal, Woolwich, to whom we wrote, and received the following reply :—

"I have then to report that there is but one sample in my hands, and that apparently a very important one—a copper instrument (a sort of knife-blade) brought from Tel Sefr by the late Mr Loftus. It is essentially copper, containing only small quantities of the impurities usually occurring in crude samples of that metal, but *no tin.* The specimen is pretty deeply incrusted with red oxide (innermost), with green carbonate, and also has considerable patches of crystallised *atacamite,* this latter resulting no doubt from the corrosion of the copper under a soil more or less charged with attractive chlorides."

That vast quantities of copper were in common use, and applied to various purposes in remote ages, is further corroborated by other writers and travellers. The author of "Notes of a Rough Ride from the East," says, respecting the ruins of ancient Persepolis—"There stood in stately solitude the pride of ages which appear almost fabulous from their distance—of empires, nearer by five centuries to the time of Noah than to ours, and of which no traces remain ; but here are sufficient to verify the narrative of their splendid existence, and to show that in some arts, and these amongst the noblest, our vaunted march of intellect is but an idle boast. Indeed, were it not from the models—the more servilely the better—this march would most certainly be a countermarch. As it is, where is the modern city that will have such a glorious wreck as this after its ephemeral, though perhaps more utilitarian, existence has passed away. Diodorus Siculus says a triple wall surrounded the place. The first wall was sixteen cubits in height, defended by parapets and flanked with towers. The second was in form like the first, but twice its elevation. The third was a square, and cut in the mountain, being sixty cubits in height, and defended by palisades of copper, and has doors of the same metal twenty cubits high." Taking twenty-one inches as the length of the cubit, these doors would be thirty-five feet high; whether cast or hammered, these are indicative of high metallurgical skill.

TIN.

TIN is very seldom found in the metallic state, and then only in very small quantities. It is abundantly found in some localities in a mineralised state, combined with oxygen and sulphur, termed respectively oxide and sulphide of tin, which constitute the ores of that metal. These ores are generally very heavy, and are found either in little pieces in the beds, or what has formerly been the beds, of rivers, mixed with sand, or in masses in veins of rocks, mixed and combined with a stony matrix. The former kind, mixed with a large quantity of earthy matter, is collected, and washed with a strong stream to free it from the impurities. The metal obtained from this is the purest, and is known in commerce as STREAM TIN. When found in veins of rocks, the ore is crushed to a fine powder, and washed in the same manner. The ore, by reason of its gravity, subsides, and is collected and taken to the smelting-house, where it is first roasted in a reverberatory furnace, to free it from the sulpher, and also from arsenic, with which it is sometimes in combination. It is then mixed with lime and carbonaceous matters, and fused for about eight hours. The carbon combines with the oxygen, and passes off as carbonic acid; while the lime unites with the silica and other earthy impurities, forming a slag. The tin is still further purified by re-fusion. When other metals, such as copper, iron, &c., are in the ore, and are reduced with the tin, the pigs or blocks of the mixture are placed in a furnace, and kept at a heat sufficient to melt tin, but not the alloyed metals. By this method the tin is sweated out, and obtained nearly pure. Sometimes the impure alloy is kept in a melted state for a length of time, when the impurities separate by gravity. The tin is afterwards subjected to operations of refining somewhat similar to copper, by melting and inserting into the mass poles of green wood.

These processes of separating and refining are so simple, that although no record of the ancient methods have been left us, we may safely conclude that the same methods obtained then as now, the only difference being the shape and construction of furnaces.

The ores of tin are not very extensively diffused through nature, although found in great abundance in a few localities. "The most remarkable feature in tin mining," says Dr Smith in his "Cassiterides," "seems to be the enduring character of the mines. Wherever tin has been produced in any considerable quantity within the range of authentic history, there it is still abundantly found. In Banca, we are told, the supply is inexhaustible; and Cornwall can

now supply as large a quantity annually as it ever
did." There is no evidence of tin having been
found in or near Egypt or Palestine, and, as already
stated by Mr Loftus, there is none in the Persian
mountains; so where the early inhabitants of
such countries obtained their tin is still a matter
of inquiry. Some authors consider that it was
obtained from the Indies; but Pliny asserts that
in his day tin was not procured from India. Some
suppose that the Egyptians obtained tin as the
Phœnicians did, from Spain and Cornwall; while
not a few consider that they obtained it through
the Phœnicians or Sidonians. That they did obtain
it from the Phœnicians at an early period of their
history is evident; but we think Egypt had sources
more direct for obtaining the great quantities they
evidently required. That tin as a separate metal
was known in very ancient times is questioned by
several able commentators on Scripture, because
the word *Bedil*, which is generally translated tin,
was equally applicable to a mixture of lead and
tin, or a sort of refuse obtained from the refining
of silver. " After an accurate investigation," says
Beckmann in his " History of Inventions," " should
everything said by the ancients of their supposed
tin be as applicable to a metallic mixture as to
our tin, my assertion, that it is probable, but by
no means certain, that the ancients were acquainted
with our tin, will be fully justified." And again
he says—" The Greek translators considered *Bedil*

to be what they called *Cassiteros*, and as the moderns translated this by *Stannum*, these words have thus found their way into the Latin, German, and other versions of the Hebrew Scriptures, which therefore can contribute very little towards the history of this metal. The examination of the word *Cassiteros* would be of more importance; but before I proceed to it, I shall make some observations on what the ancients called *Stannum*. This at present is the general name of our tin, and from it seems to be formed the *estain* of the French, the tin of the low German and English, and the *znin* of the high German. It can, however, be fully proved that the stannum of the ancients was no peculiar metal, at any rate not our tin, but rather a mixture of other metals, which, like our brass, was made into various articles, and employed for different purposes, on which account a great trade was carried on with it."

"The oldest mention of tin is in Scripture— Numbers xxxi. 22. Moses seems to name all the metals then known; and besides gold, silver, copper, iron, and lead, he mentions also *Bedil*, which all commentators and dictionaries make to be tin. It seems probable that in this passage *Bedil* is our tin; but must it not appear astonishing that the Midianites in the time of Moses should have possessed this metal?"

We do not feel the same astonishment at tin

being named, even in this early age, along with copper and iron—metals more difficult to obtain; neither do we think Beckmann need feel, consistently, surprised at the existence of this metal in these times, as he himself admits that bronze was known from the very earliest ages—and this being a combination of tin and copper, it could not have been made without tin; or if the word applied to a mixture of lead and tin, from which bronze was made, then such bronze would yield, by analysis, lead, copper, and tin, a supposition which the analyses of ancient bronze do not verify. The circumstance that the same word was applied to a mixture of lead, tin, and silver, as well as to pure tin, does not affect the probability of pure tin being known. Lead and tin are very seldom found together in the same ore; and where they are found mixed, it has been the result of accident or art. In our own workshops, it is quite common to call the technical term solder, *tin*, although known to have much lead in it. How obscure, then, would any description of art be to the general reader were technical terms used without explanation; and may not this be the source of much of the obscurity in Pliny, and others who treat of practical arts, while they were not themselves skilled in them? As illustrative of such obscurity, take the following translation from Pliny :—

" Black lead has a double origin, for it is either produced in a vein of its own, without any other

metal, or otherwise it is mingled with silver in
the same mine, being mixed together in the same
stone of ore, and they are only separated by melting and refining in a furnace. The first liquor
that flows from the furnace is tin (stannum), and
the second silver; that part which remains behind
is galena, the third element of the vein, which
being again melted, after two parts of it are
deducted, yields black lead." The first part of this
quotation, which concerns the natural production
of the ore, is clear, and easily understood; the
second part, descriptive of the practical process,
is as obscure as could well be conceived. We
quote the following from an interesting paper
by J. Phillips, F.R.S., on ancient mining in
Britain :—

" Tin, the ore of which has been found at the
surface in many situations, with auriferous sand
and gravel, cannot have been long unknown to
the gold-finders in the East and West. Some one
of the many accidents which may, and rather must,
have accompanied the melting of gold, would
disclose the nature of the accompanying metal,
whose brilliance, ductility, and very easy fusibility, would soon give it value. The melting of
tin ore is, however, a step in advance of the fusing
of native gold. The gold was fused in a crucible
made of white clay, which only could stand the
heat and chemical actions which that generated;
but tin ore would, in this way of operation, prove

entirely infusible.   It must be exposed at once to
heat, and a free carbonaceous element.   The easiest
way of managing this is to try it on the open hearth.
Perhaps some accidental fire in the half-buried
bivouacs of the Damnonii may have yielded the
precious secret.   As to the fuel, we are told that
pine wood was best for brass and iron; but the
Egyptian papyrus was also used, and straw was
the approved fuel for gold.   In the metalliferous
county of Cornwall peat is plentiful.

"As the bellows were used at least a thousand
years before Pliny, we have here all the materials
for a successful tin-smelter's hearth.   If the smelting works were on waste land, and a little sunk in
the ground, we recognise the old 'bole' or 'bloom-
ing' of Derbyshire, now only a traditional furnace,
but anciently the only one for the lead and iron of
that county.   Pure tin once obtained, there must
intervene a long series of trials and errors before
its effect in combination with lead, brass, and
silver, could be known—before the mode of conquering the tendency to rust in the act of soldering can be discovered.   From all this it follows
that the smelting of tin might be, and probably
was, performed by the inhabitants of the Cornish
peninsula.   This art they may have brought from
the far East—Phœnicians may have taught it
them.   But all the accounts of the ancient tin
trade represent the metal, and not the ore, as being
carried away from the Cassiterides.   Diodorus

mentions the weight and cubical form of the tin in blocks carried from Ictias to Marseilles and Narbonne, and Pliny says of the Gallecian tin that it was melted on the spot."

Dr Alexander, in his "Ancient British Church," quotes from the Rev. F. Thackeray's "Researches" the following passage :—"The tin, lead, and skins of Britain, instead of being immediately shipped in Cornwall, or other maritime districts, are said to have been taken to the Isle of Wight, thence transported to Vannes, and other ports of Brittany, afterwards conveyed overland to Marseilles, and finally exported to all parts of the world which traded with the Greeks." Dr A. also states that Bochart deduces the word *Britannia* from the Hebrew word *Barab-anac,* "the land of tin."

Mr Richard Edmonds discovered some fragments of a bronze furnace in Cornwall, which he considers had been brought there by the Phœnicians, who gave such things in exchange for lead, tin, and hides. After some observations upon the locality where these fragments were found, and other correlative circumstances, he considers the manufacture of these fragments to date back not less than two thousand years. They were coloured by charcoal, and had a deep-green coloured patna upon them. The analysis of these by Mr R. Hunt gave—

Copper,	.	.	.	.	.	72
Tin, .	.	.	.	.	.	9
Iron, .	.	.	.	.	.	4
Earthy matters,		.	.	.	.	3
Carbonic acid, &c.,		.	.	.	.	12
						100

The town where these relics were found is the oldest in the country, and has two names— *Market Jew*, and *Marazion* or *Marghasion*, which means in Hebrew Market Mount. Mr Edwards therefore considers that Jews had found their way in very ancient times to this, and had purchased the ores of tin from the natives, and smelted them, and sold the metal to the Phœnicians; and the name of the locality was therefore called by the Jews Marazion, and by the natives Market Jew, hence the double name. If the Jews ever established a market in Cornwall to which the natives could bring their ore, and also set up smelting works to prepare the tin for the Phœnicians, it must have been at a comparatively recent time after the Phœnicians had taught the Jews the way to Cornwall; but more probably the Jews who were located there were captives taken by the Phœnicians, and put to work in their smelting houses. The remnants of smelting pits found in Cornwall, says Mr Edwards, are termed *Jew's houses*.

Dr Daniel Wilson, in his "Prehistoric Annals of Scotland," speaks thus of tin :—

"The familiarity of the ancient Britons with tin,

though this metal does not occur in a native state,
may be readily accounted for from the ore being
frequently found near the surface, and requiring
only the use of charcoal, and a very moderate
degree of heat, to reduce it to the state of a metal.
We have no specific mention of any other source
from whence the ancients derived the tin which
they compounded with the copper found so abundantly in several parts of Asia; with the single
and somewhat vague exception made by Strabo,
when he calls a certain place in the county of the
Drangi, in Asia, by the name of Cassiteron. That
tin was known, however, from very early times,
not only by the discovery of numerous early
Egyptian and Assyrian bronze relics, but also by
its being noted by Moses among the spoils of the
Midianites, which were to be purified by fire;
and by Ezekiel, among the metals of which Tarshish was the merchant of Tyre. . . . The
allusions of Herodotus leave no room to doubt that
his information was derived indirectly from others.
The Phœnicians long concealed the situation of
the Cassiterides from all other nations; and even
Pliny treats as a fable the report of certain islands
existing in the Atlantic, from whence white lead or
tin was brought. It need not, therefore, surprise
us to learn so little of these islands from ancient
writers, even though we adopt the opinion that
they continued for many centuries to be the chief
source of one of the most useful metals. Antimony

is found in the Kurdish mountains, and pure
copper ore abounds there as well as in the desert
of Mount Sinai; but no tin is found throughout
any part of Assyria. It is a metal of rare occur-
rence, though found in apparently inexhaustible
quantities in a very few localities. The only
districts, according to Berzilius, where it is
obtained in Asia, are the islands of Banca, only
discovered in 1710; and the peninsula of Malacca,
where Wilkinson conceives it possible that tin may
have been wrought by the Egyptians. The mines
of Malacca are very productive, and may have been
the source from whence Tyre derived ' the multitude of riches; ' but we have no evidence in
support of such conjectures. Cornwall still yields
a larger quantity of the ore than any other locality
in the old or new world where it has yet been
discovered, and many thousands of tons have been
exported by the modern traders to India, China,
and to America. Taking all these circumstances
into consideration, it seems in no degree improbable that long before Solomon sent to Tyre for a
' worker filled with wisdom and understanding, and
cunning to work all works in brass,' or employed the
fleets of Hiram, king of Tyre, to bring him precious
metals and costly stones for the temple at Jerusalem, the Phœnician ships had passed beyond the
pillars of Hercules, and were familiar with the
inexhaustible stores of these remote islands of the
sea, which first dawn on the history as the source

of this most ancient alloy. Strabo's description of the natives of the Cassiterides is not to be greatly relied upon. According to him they were a nomade, pastoral race, of peaceful and industrious habits ; but he refers especially to their mines of tin and lead, the produce of which they exchanged with the foreign traders, along with furs and skins, for earthenware, salt, and copper vessels and implements."

Since Drs Alexander, Wilson, and Mr Philips wrote, Dr George Smith has made a very exhaustive inquiry into the source of tin in ancient times. Concerning the statement that the metal was taken to the Isle of Wight, and thence to Marseilles, he writes—" It is founded on a great error in commercial geography. Diodorus simply states that the men who dig and prepare the tin carry it to a British isle near at hand." This, Dr Smith argues, cannot mean the Isle of Wight, which is two hundred miles distant; and then he says—" The trade of which we speak began at least as early as 1200 B.C., and lasted more than a thousand years. Marseilles was not founded earlier than 600 B.C., and Narbo not until 400 years later. It is presumed that none will contend that tin was taken from Britain by this overland route before that date—600 B.C. If the Phœnicians conducted their commerce in Cornwall, it would be infinitely more to their advantage to take the commodity direct to their vessels at Cadiz, than to transport it to Gaul, allow it to go entirely out of their hands for a thirty-days' land journey, and then to re-ship it at the mouth of the Rhone. This could not have been the case; yet prior to this date the trade had attained its widest extent."

About the year Marseilles was founded, Ezekiel speaks of the commerce of Tyre, and names tin procured from Tarshish as one of the chief commodities sold in their fairs. All our researches have conducted us to the conclusion that the principal part of this tin was found in Britain, and brought from thence in Phœnician vessels, by the way of Gades to Tyre. The maritime traffic of the Phœnicians is of very early date. The city of Tyre was founded, according to Dr Hales, 2267 B.C.; but before Tyre there was an older city—Zidon. Jacob, about 1900 B.C., alludes to the practice of navigation by this people thus— "Zebulon shall dwell at the haven of the sea, and he shall be for a haven of ships, and his border shall be unto Zidon." Joshua, about 1600 B.C., calls Zidon " the great Zidon," and speaks of the more recent capital as " the strong city of Tyre," and there is every evidence of their early excellence in manufactures. " The vase of silver which Achilles proposes as a prize in the funeral games in honour of Patroclus was a work of the skilful Sidonians. The garment offered by Hecuba as a propitiation to Minerva was the work of

Sidonian women, whom Paris had brought with him to Troy when he visited Phœnicia. The bowl of silver with edges of gold which Menelaus gives to Telemachus is called a work of Hephaistos, and was given to him by a king of the Sidonians."

Respecting tin Dr Smith sums up :—" We recall attention to the simple fact that tin was an article of ancient commerce, at least as early as B.C. 1200, on the eastern shores of the Mediterranean. This is an established truth. It is certain that tin was used and sold at Sidon and Tyre at this early date. Whence, then, did it come? The universal testimony of all history and tradition answers, from Britain. This testimony has been received, and the British origin of the tin supplied as an article of commerce in the earliest times has been believed by great numbers of learned men in different ages and countries. Having carefully studied the subject, they have been fully convinced that the ancient Phœnicians traded with Britain for this metal, and regularly took it from the coast of this island, in Phœnician ships, to Tyre and Sidon. The names of those who have entertained this opinion would, if collected, exhibit a body as numerous, as intelligent, and as entitled to deference and respect, as could be found supporting almost any historical truth."

It is very probable that in ancient times the tin ore was smelted at or near the mines, or in

the country where the mines were situated, and
only the metal sent away in ships. Hence we
have no allusions to tin smelting operations
in Scripture; but Diodorus, speaking of the
people of Britain, says—" They that inhabit the
British promontory of Balerium (Land's End), by
reason of their converse with strangers, are more
civilised and courteous to strangers than the rest
are. These are the people that prepare the tin,
which with a great deal of care and labour they
dig out of the ground; and that being rocky, the
metal is mixed with some veins of earth, out of
which they melt the metal, and then refine it.
Then they cast it into regular blocks, and carry it
to a certain place near at hand called Ictis."

The first mention of tin in the Scriptures is
in the time of Moses, after the Israelites had
fought against the Midianites. It constituted a
part of the spoils taken from that people, and the
reference is to the method by which such metals
were to be made ceremonially pure, thus—" Only
the gold and the silver, the brass, the iron, the
tin, and the lead, everything that may abide the
fire, ye shall make it go through the fire."

This ceremony of passing the metals through
the fire has been supposed by some to mean
making them red-hot. Such an idea is absurd.
Many of the metallic articles taken in this and
other wars were no doubt manufactured, and the
attempt to make them red-hot would destroy

them. Tin, for instance, melts at a temperature
of 420° Fahr., about twice the heat of boiling
water, and only half of the temperature of redness; and lead melts at 620° Fahr. : so that if the
passing through the fire meant making them red
hot, it would melt both the tin and the lead. The
passing through the fire, in our opinion, means a
mere heating, to destroy organic matter, which
would be effected by passing them *suddenly*
through a fire or flame. We have seen this
method practised before *bath-brick*, or such polish-
ing materials, had reached remote dwellings—
the blades and prongs of knives and forks, before
being used, were plunged into the red, glowing
embers, and then wiped clean and made ready for
use. The same practice is referred to by the
prophet Ezekiel—"Then set it empty upon the
coals thereof, that the brass of it may be hot, and
may burn, and that the filthiness of it may be
molten in it, that the scum of it may be consumed."

The relative value of tin compared with the
other ordinary metals is nowhere stated, but
that it was classed with the ordinary, or inferior
metals, is apparent from various passages in
Scripture, and other ancient writings.

There are no indications that tin was used alone
in the manufacture of vessels or ornaments;
although the ease with which it melts and can be
cast, the whiteness of its colour, and its not
being subject to tarnish, incline us to the belief
that it was used alone for various purposes.

Neither is there reference in Scripture to the use of tin for overlaying or plating other metals. Pliny, however, speaks of tin-plate and the coating of copper utensils with tin as being common in ancient times, and as having been done in a very perfect manner. He says the articles so tinned could scarcely be distinguished from silver, and adds that the tinning did not increase the weight of the vessels, the metal being applied so thin.

Beckmann says—"That vessels were made of cast tin at an early period is highly probable, although I do not remember to have seen any of them in collections. I am acquainted only with two instances of their being found—both of which occurred in England. In the beginning of last century, some pieces of tin were discovered in Yorkshire, together with other Roman antiquities; and in 1756 some tin vessels of Roman workmanship, with Roman inscriptions, were dug up in Cornwall."

From the ease with which tinned iron is destroyed, by exposure to a damp atmosphere, it could hardly be expected to be found now, although the process had been in general use in ancient times.

Other applications of tin, which give evidence of great advancement in the arts, are known to have existed in very early times. According to recent discoveries, the oxide of tin was used, both by the ancient Egyptians and Assyrians, for enamelling and glazing earthenware, an application which until lately was considered quite a modern invention.

In support of the supposition that tin in ancient times was occasionally used by itself for objects of art, we have an eminent corroboration in that classical puzzle to the learned—the shield of Achilles. This famous shield is said to have been made of five plates of metal, two of which were tin. The following is a literal translation of the passage in the "Iliad," describing the making of the shield of Achilles, translated by a friend of ours, who is an eminent Greek scholar :—

"Thus speaking, he indeed left her there, and went to the bellows. These indeed he turned to the fire, and ordered them to labour. And twenty pair of bellows all blew in the melting pit, sending forth a manifold vehement blast of air—at one time indeed to assist him to hurry on the work, at another time the contrary, just as Vulcan might wish, and the work might be finished. And he threw into the fire indestructible brass and tin, and precious gold and silver. Then afterwards he placed on the stithy a huge anvil, and he grasped in his hand a powerful hammer, and with the other he grasped a pair of tongs.

"And first of all he made a shield, both long and strong (closely pressed), working it curiously on all sides; and he surrounded it with a glittering rim, threefold, flashing, and from it a silver-studded strap. And indeed there were five plates (layers) of the shield itself, and in it he made many curiously wrought things with knowing mind."

We refer to this extraordinary piece of art here,

because it is one of the few instances where tin is
mentioned as used alone. The use of tin in this
way has caused some commentators on Homer to
write with considerable confusion, and to deduce
erroneous deductions. The latest instance we have
seen is that of the Right Honourable W. E. Gladstone, in his "Juventus Mundi." We take the
liberty of quoting several passages from his book—

"Archæological inquiry is now teaching us to
investigate and to mark off the periods of human
progress, among other methods, by the materials
employed from age to age for making utensils and
instruments. And the poems of Homer have this
amongst their many peculiarities; they exhibit
to us, with as much clearness, perhaps, as any
archæological investigation, one of the metallic
ages. It is, moreover, the first and oldest of the
metallic ages, the age of copper, which precedes
the general knowledge of the art of fusing
metals, which (as far as general rules can be laid
down) immediately follows the age of stone, and
which in its turn is probably often followed by the
age of bronze, when the combination of copper with
tin has come within the resources of human art.

"The grand metallic operation of the poem is
that of Hephaistos in the production of the
shield. The metals used were gold, silver, tin,
and *chalcos*—which has been, by mere licence of
translators, interpreted as brass, for there was no
brass till long ages after Homer had rolled away—
which has been more plausibly taken to mean

bronze—and which, after a good deal of inquiry, I am satisfied can only mean copper, either universally or absolutely, or as a general rule with very insignificant exceptions.

"The discussion would be too long for this place; but the passage immediately before us of itself affords almost sufficient instruction.

"In the formation of the shield there is no mixture or fusion of metals. The same, and all the same which are put into the roaring fire, reappear each by its original name in various portions of the shield. There is indeed one passage where a trench is represented, and this is called *kuanèe*, a word meaning either made of *kuanos*, or like *kuanos* in colour. There are two reasons for giving the latter signification to the word: one that it commonly bears that sense in Homer; the other that although *kuanos* may have been a mixed metal, yet there is no sign of founding or casting in the great masterpiece of Hephaistos.

"He could only mix by melting; and had he melted metals we must have heard of moulds to receive them. Instead of this, the only instruments which he makes ready for the work are—

"1. The anvil;

"2. The hammer in (right) hand;

"3. The pincers in his left.

"It is plain, then, that he was supposed not to melt, but only to soften the metals by heating, and then to beat them into the form he wished to produce.

" Had Homer been conversant with the fusing or casting of metals, this is the very place where we must have become aware of it; especially as his works of skilled art are all of Phœnician origin or kin, and his Hephaistos is a god of Phœnician associations.

" If *chalcos* be not copper, then copper is never mentioned in Homer. But in an early stage of society copper was commonly by far the cheapest and most accessible of metals, and it is quite impossible to suppose that we never once hear of copper from an author who incessantly makes mention (so it is argued) of another metal, whereof it is by far the largest component part."

Against these conclusions we find several objections. In the first place, where and how was the plentiful supply of copper obtained? for the author says, "This was an age preceding the knowledge of fusing metals."

Was all the copper before and in Homer's time found in the metallic state, and so pure that it could be hammered and worked into tools and implements for general use. We have already shown that methods of melting and casting this metal were known centuries before Homer; and although considerable masses of metallic copper are sometimes found, yet such finds are rare, and could not make it cheap and common, especially in an age when metals were abundantly used. "Copper," says Job, "is molten out of the stone," and we have no doubt this was also the method of obtaining it in

Homer's days, and which made it "the only one
metal common in Greece at that time."

In the second place, Mr Gladstone says of the
making of the shield, "The same, and all the
same which are put into the roaring fire, reappears
each by its original name," &c. As mentioned
before, tin melts at 420° Fahr., far under the temperature of red heat. What the result would be of
putting this metal into a roaring fire is easily conceived: neither anvil, hammer, nor pincers would
be of much use for it. Then "the roaring fire was
kept up by a blast of twenty pairs of bellows."
Whether the fuel was wood or charcoal, copper,
gold, and silver would all melt in such a fire with
ease. Our own opinion is, that the description of
the making of the shield is not to be taken as a
correct technical detail of the process, of which in
all probability the poet was ignorant, but that he
simply drew a poetic figure from popular knowledge, and it being made by divine aid, a large
licence was taken. We shall be able to form
a more correct judgment from the description
given of the shield after it was finished, which
finished article the poet probably had seen, or had
it described to him. The passage is thus translated by our friend—"Nor did then the mighty
spear of the warlike Augas break through the
shield, for the gold, the gift of a god, hindered it
(kept it back), but indeed it drove (pierced)
through two folds; but verily there were three
remaining, since Vulcan had beaten out five plates,

two of brass, and two within (behind) of tin, and one golden (viz., two front of brass, two behind of tin, and the middle of gold), by which, indeed, the brazen spear was held." Another translation of the same passage runs thus—

> "Of five plies it piercèd only two,
> And these were brass; there still remainèd three,
> The one of gold, the other two of tin;
> And stopped by the brass it could not be,
> But coming to the gold, it thus stuck in."

The word translated *plies* here is by some translators called plates. This is all clear and distinct.

Of Vulcan's use of the bellows we find in another part of the poem these words—"That fixed his coals sometimes with soft and sometimes with vehement air, as he willed, and his work required." That is, as the metal in the fire at the time required more or less heat to melt. We think that they were all melted, and then cast in the same mould separately; and these castings Vulcan took with his pincers and hammered, smoothing and hardening the metals, and fitting them for binding together with the silver-studded strap.

Mr Gladstone says the word *kuanèe* or *kuano* in Homer "generally means the type of a very dark colour, and may possibly mean bronze. The Greeks had bronze in small quantities; it was more valuable than copper, but apparently less prized than tin. In the planetary worship of the

East six deities were connected with six pure metals, and one with *kuanos*. In Homer we find the six metals with *kuanos*. Now, as the septiform system was apparently represented in the seven gates of Thebes, and as the Greeks evidently depended on the Phœnicians for imported metals, I conclude that *kuanos* was the seventh metal, a mixed one, and I know no conclusive reason why it should not be bronze." Although this conclusion is opposed to some of his former statements, it nevertheless gives the key to Hephaistos's operations, and to Mr Gladstone's difficulty respecting the word *kuanos*. The word translated trench by him, and pit by others, we believe refers to the mould in which the different plates of the shield were cast; and this mould was made of bronze, and such a mould would certainly be aptly typified, or would typify a dark colour, as such moulds when used become very dark. That bronze moulds were used in very early times has been clearly proven by such being discovered in ancient ruins.

Concerning what Mr Gladstone says about bronze, we may have something further to say in the following chapter; but his inference that society was at the time of Homer only in what is termed the copper age, the age following stone and preceding bronze, is certainly most extraordinary in the face of history and archæological researches.

K

Since the above went to press, Mr Gladstone has published a poetic translation of Homer's description of the manufacture of the shield, from which we copy the following :—

I.

So he spake, and left the goddess ;
Straightway to the bellows drew,
Fixed them fireward, set them blowing,
Mouths a score in all they blew—
Reddening, whitening all the furnace
With their timely various blast,
As the god and work required it,
Slower now, and now more fast.
Precious gold and stubborn copper,
Silver store and tin he cast
In the flame.   The ponderous anvil
Next upon its block he tries ;
One hand grasps the sturdy hammer,
One the pincers firmly plies.

II.

First of all the shield he moulded,
Broad, and strong, and wrought throughout
With a bright and starry border
Threefold thick, set round about.
Downward hung its belt of silver,
Five the layers of the shield,
And with skilful mind he sculptured
Rare devices o'er its field.

In his preface to this poem, Mr Gladstone says, "The materials used in the composition of the shield deserve notice. The metals cast into the furnace are copper, tin, gold, and silver." If copper was put into the furnace, it entirely disappears in the shield, which was composed of two plates bronze, two of tin, and one of gold. So that

one of two things must follow: either Hephaistos
made bronze of the copper when making the shield,
or Mr Gladstone's translation of the word *copper*
is wrong. One idea in the lines above suggests
an answer as follows—

> " Precious gold and stubborn copper,
> Silver store and tin he cast
> In the flame."

There is no practical sense in which *stubborn* can
be applied to copper when compared with gold and
silver. Copper melts at a lower heat than gold, and
it is as malleable as silver; but if *bronze* is substituted for copper, then *stubborn* is appropriate, as
bronze is much less malleable than either silver or
gold. We are therefore strengthened in our opinion
that Mr Gladstone is wrong when he says in his
" Juventus Mundi " that " in the formation of the
shield there is no mixture or fusion of metals."
In the above poem there are many indications of
castings which could not be made without fusion,
such as—

> " There a herd of kine he moulded,
> Some in tin and some in gold,"

which were no doubt cast and fastened upon the
shield after it was made.

BRONZE.

AMONG the great historic nations of antiquity, the chief consumpt of copper and tin was in the manufacture of bronze; and the quantities of these metals necessary for this purpose must have been very great, for bronze seems to have been the principal metallic substance of which articles both of utility and. art were formed. Wilkinson, Layard, and others, found bronze articles in abundance amongst the *débris* of all the ancient civilisations to which their researches extend, proving that the manufacture of this alloy was widely known at a very early period; and strange to say, when we consider the applications of some of the tools found, we are forced to the conclusion that the bronze of which they were made must originally have been in certain important particulars superior to any which we can produce at the present day. In these researches were found carpenters' and masons' tools, such as saws, chisels, hammers, &c., and also knives, daggers, swords, and other instruments which require both a fine hard edge and elasticity. Were we to make such tools now, they would be useless for the purpose to which the ancients applied them. Wilkinson says, "No one who has tried to perforate or cut a block of Egyptian granite will scruple to acknowledge that our best steel tools are turned in a very short time, and require to be re-tempered; and the labour experienced by the French engineers who removed the obelisk of Luxor from Thebes, in cutting a space less than two feet deep along the face of its partially decomposed pedestal, suffices to show that, even with our excellent modern implements, we find considerable difficulty in dóing what to the Egyptians would have been one of the least arduous tasks."

But Wilkinson believes that bronze chisels were used for cutting granite, as he found one at Thebes, of which he says, "Its point is instantly turned by striking it against the very stone it was used to cut; and yet, when found, the summit was turned over by blows it had received from the mallet, while the point was intact, as if it had recently left the hands of the smith who made it."

"Another remarkable feature in their bronze," says the same author, "is the resistance it offers to the effects of the atmosphere—some continuing smooth and bright though buried for ages, and since exposed to the damp European climate. They had also the secret of covering the surface with a rich patina of dark or light green, or other colour, by applying acids to it."

It is probable that many of the children of Israel,

while slaves in Egypt, were workers in foundries and other works where metallic articles were manufactured. If this conjecture be correct, it will help to explain that interesting portion of Scripture history, the journeyings in the wilderness, and account for the ability they possessed of working in metals. Their casting and working into shape such articles as the calf, the serpent, and the more elaborate metallic work of the Tabernacle, is strong circumstantial evidence in favour of the supposition that the Israelites while in Egypt were not as a whole employed in the labour of brickmaking.

But before entering on an investigation of the bronze works of art mentioned in the Bible, let us direct our attention to the manufacture of bronze as it is carried on at the present day. Bronze, then, is a mixture of copper and tin, and by varying the proportions of mixture of these metals different qualities of bronze are produced. One quality may be hard but brittle, another both hard and flexible; a third when polished may yield a bright, reflecting surface, suitable for mirrors; while a fourth may be found excellent bell metal. To make good bronze requires very careful attention during the operation of mixing the two metals. Something more is required than merely fusing a certain quantity of copper and tin together; or, as is generally supposed to be the case, fusing copper, and adding some tin to it. To obtain a good homogeneous alloy, hard and tough, the tin ought to be mixed with due proportions of copper, and cast into blocks, and this 'alloy remixed with more copper, and that probably several times, if there be no old bronze to add to the mixture. Copper and tin which have impurities in them do not make good bronze for tools, swords, daggers, guns, &c., but bronze for coins, for casting figures, and such like, does not require to be so pure.

Again, the tendency, on account of the different gravities of the two metals, when fused together to partially separate and settle in strata, affects greatly the quality of the bronze; and it requires great skill and watchfulness to avoid this, by casting at a particular heat, and also by poling—that is, by inserting poles of green wood into the melted mass, which not only causes violent ebullition, but prevents oxidation of the metals. A knowledge of these processes and conditions being necessary in order to make good bronze, such knowledge must have been possessed by the ancients before they could produce such bronze as we find they did make. The acquirement of such knowledge must have been a gradual process stretching over a long period, and society must have been in a very advanced state when such exact proportions, processes, and conditions were so well understood that different qualities could be produced at will.

Very few analyses of ancient bronze are recorded compared with what might have been expected, considering the number of ancient articles in bronze which we now possess, and the interesting nature of the inquiry. The few which we have noticed, however, are worthy of consideration.

We have already referred to a chisel which Wilkinson found in an ancient quarry in Egypt. This, when analysed, gave—

Copper, ...	. 94·0
Tin, ...	. 5·9
Iron, ...	. ·1
	100·0

An ancient chisel from Peru, analysed by Vauqueline, gave—

Copper, . . .	. 94
Tin, ...	. 6
	100

These two appear to be exactly the same quality of alloy. The density of the last is given as 8.815, which is very high; the density of Wilkinson's chisel is not given. Bronze varies considerably in density, even when the metals are mixed in the same proportions. The purer the metals, the greater the density. We recollect a statement made by the late Colonel Dundas, who at the time was chief of the arsenal department, concerning a bronze cannon made from very pure copper; that although the proportional composition

of the alloy was the same as that of the other
cannons in the arsenal, its density was greater
than any they had ever had.

A dagger (locality not given), analysed by
Klaproth, gave—

Copper,	. . . .	. 91·6
Tin,	. . . .	. 7·5
Lead, .	. . . .	. ·9
		100·0

A dagger (locality not given), analysed by
Heslm, gave—

Copper,	. . . .	. 93·4
Tin, .	. . . .	. 6·6
		100·0

An Egyptian dagger, analysed by Vauqueline,
gave—

Copper,	. . . .	. 85·0
Tin, .	. . . .	. 14·0
Iron, .	. . . .	. 1·0
		100·0

Celtic arms, analysed by Fresenius, gave—

Copper,	. . . .	. 92·00
Tin, .	. . . .	. 6·70
Lead, .	. . . .	. ·69
Iron, .	. . . .	. ·29
Nickel,	. . . .	. ·31
		99·99

Gallic axe, analysed by Girardine, gave—

Copper,	. . . .	. 85·90
Tin, .	. . . .	. 14·10
		100·00

Among ancient bronzes the very ancient is generally found to be the purest in quality, that is, has less mixture of other metals, and we believe if more attention were paid to the composition of ancient bronzes, the localities where found, and probable dates of manufacture, we might yet trace the sources where the metals, the copper especially, were obtained. When we find bronze containing cobalt and nickel, it is highly probable that these metals were in the copper, and that the copper was obtained from localities where such metals are found. Copper is found abundantly where no cobalt or nickel are known to exist.

In Moore's " Ancient Metallurgy " we find the following analyses of very pure bronze :—

A Grecian helmet, no date given—

Copper,	.	.	.	.	.	81·5
Tin,	.	.	.	.	.	18·5
						100·0

A sword, found in the peat-moss of the Souime—

Copper,	.	.	.	.	.	87·47
Tin,	.	.	.	.	.	12·53
						100·00

A sword, found at Abbeville—

Copper,	.	.	.	.	.	85
Tin,	.	.	.	.	.	15
						100

Another sword, where found not stated—

| Copper, | . | . | . | . | 90 |
| Tin, | . | . | . | . | 10 |

100

Sword-blade, found under Chertsey Bridge, Thames. Analysed by J. A. Phillips—

Copper,	.	.	.	.	90·00
Tin, .	.	.	.	.	9·54
Iron, .	.	.	.	.	·33

99·87

Fragment of sword-blade, found in Ireland. Analysed by Phillips—

| Copper, | . | . | . | . | 91·39 |
| Tin, . | . | . | . | . | 8·38 |

99·77

Celt—

| Copper, | . | . | . | . | 90·18 |
| Tin, . | . | . | . | . | 9·82 |

100·00

The analyses in general agree with many synthetical experiments, made by Dr Pearson, with different proportions of tin and copper. He came to the conclusion that the fittest proportion for the manufacture of weapons and tools is—one part of tin to nine parts of copper. Of ancient bronzes for different purposes we add the following :—

Bowl or dish, from Nimroud. Analysed by Dr Percy—

| Copper, | . | . | . | . | 87·57 |
| Tin, . | . | . | . | . | 10·43 |

100·00

Hook, from Nimroud. Analysed by Dr Percy—

Copper,	.	.	.	.	. 89·85
Tin,	.	.	.	.	. 9·78

99·63

Bronze, overlaying iron. Analysed by Dr Percy—

Copper,	.	.	·.	.	. 88·37
Tin,	.	.	.	.	. 11·33

99·70

"This," says Dr Percy, "was a small casting in the shape of the foreleg of a bull; it formed the foot of a stand, consisting of a ring of iron resting upon three feet of bronze. It was deeply corroded in places, and posteriorly was fissured at the upper part. A section was made, which disclosed a central piece of iron, over which the bronze had been cast. The casting was sound, and the contact perfect between the iron and surrounding bronze. It was evident, on inspection, that the bronze had been cast round the iron, and that the iron had not been let into the bronze. Some interesting considerations are suggested by this specimen.

"The iron was employed either to economise the bronze for the purpose of ornament, or because it was required in the construction. If the former, iron must have been much cheaper than bronze, and therefore, probably more abundant than has been generally supposed. No satisfactory conclusion can be arrived at on this point, from the fact

that bronze antiquities are much more frequently
found than those of iron, for the obvious reason
that bronze resists much better than iron destruction by oxidation. Although I think there
are reasons to suppose that iron was much more
extensively used by the ancients than seems to be
generally admitted, yet in the specimen in
question it appears to me most probable that the
iron was used because it was required in the construction. And if this be true, the Assyrians
teach a lesson to many of our modern architects
and others, who certainly do not always employ
metals *in accordance with their special properties.*
The instrument under consideration, it will be
borne in mind, was one of the feet of a stand,
composed of an iron ring, resting upon vertical
legs of bronze. A stand of this kind must have
been designed to support weight, probably a large
caldron; and it is plain that the ring portion
should therefore be made of the metal having the
greatest *tenacity*, and the legs of metal adapted to
sustain *vertical or superior cumbent weight.* Now,
this combination of iron and bronze exactly fulfils
the conditions required. I do not say that a ring
of bronze might not have been made sufficiently
strong to answer the purpose of the ring of iron;
but I do say that, in that part of the instrument, iron is more fitly employed than bronze.
Moreover, the contrast of the two metals may have
been regarded as ornamental."

Bell. Analysed by Dr Percy—

Copper, ...	84·70
Tin, ...	14·10
	98·80

Here it will be observed that where sound is required the proportion of tin is increased.

Dr Percy also remarks upon the general character of the metallic objects obtained by Mr Layard:— "There are numerous other objects in metal in Mr Layard's collection at the British Museum which are extremely interesting in respect to the mode of manufacture, and which require very accurate examination before they can be properly described by the metallurgists. The beautiful workmanship of the vessels which Mr Layard believes to have been used in religious ceremonials is especially deserving of attention, and demonstrates the skill of the Assyrians in their treatment of bronze.

"One specimen particularly deserves attention. It was a thin, hollow casting in bronze, which was attached to the end of one of the arms of the throne. The casting had evidently been chased, and for that purpose, must have been filled with some soft material, such as pitch, which is used at the present time. In the interior there was some black matter, which, on examination, I found to burn like pitch, and leave an earthy residue, so that probably a mixture of asphaltum and earth had been employed for the purpose mentioned."

Although the remark of Dr Percy about the bell having more tin than the other bronzes, to give sonorousness, is very ingenious, and probably correct, yet in the preceding analyses we find swords and other articles having more tin compared with the copper than the Babylonian bell.

We take the following analyses from Wilson's "Prehistoric Annals of Scotland:"—

1st, A lituus, or musical wind-instrument, found in the river Witham, Lincolnshire, gave—

Copper, ...	88
Tin, ...	12
	100

2d, Spear-head, which had been cast, as was evident from its rough surface, figure, texture, and grain—

Copper,.	86
Tin, ...	14
	100

This had a little silver.

3d, A saucepan, also cast—

Copper, ...	86
Tin, ...	14
	100

4th, Scabbard: this had an iron sword within—

Copper, ...	90
Tin, ...	10
	100

Leaf-shaped sword. Analysed by Professor George Wilson—

Copper,	.	.	.	.	88·51
Tin,	.	.	.	.	9·30
Lead,	.	.	.	.	2·30
					100·11

Axe-head. By Professor Wilson—

Copper,	.	.	.	.	88·05
Tin,	.	.	.	.	11·12
Lead,	.	.	.	.	·78
					99·95

Palstave or celt. By Professor Wilson—.

Copper,	.	.	.	.	81·19
Tin,	.	.	.	.	18·31
Lead,	.	.	.	.	·75
					100·25

Bronze caldron. By Professor Wilson—

Copper,	.	.	.	.	92·80
Tin,	.	.	.	.	5·15
Lead,	.	.	.	.	1·78
					99·73

The small portions of lead found in some of these we do not think was intentionally added as a part of the alloy, but may have been an impurity either in the tin or copper, or it may have been that the manufacturers used lead in refining their copper as we do now, and a small portion remained in the refined copper, as will be found in all our tough-cake copper, that is, copper made for rolling and hammering at the present day. But this cannot be said of the following, also analysed by Dr George Wilson :—

Caldron—

Copper,	.	.	.	.	. 84·08
Tin,	.	.	.	.	. 7·19
Lead,	.	.	.	.	. 8·53
					99·80

Roman camp-kettle—

Copper,	.	.	.	.	. 88·22
Tin,	.	.	.	.	. 5·63
Lead,	.	.	.	.	. 5·88
					99·73

Here there has evidently been lead added to the alloy; but although all of them are in all probability of Roman manufacture, it may be that centuries intervene between the dates when they were made. And as we have already said, when the dates and localities of the manufacture of these relics are found, it will give us a greater insight into the progression of metallic art; or otherwise, the analyses of any ancient bronze would be a key to the date and place of manufacture.

For articles such as coins, medals, figures, and statuary, less care was taken in the proportions and purity of the metals used in the alloy, as is evidenced by a few analyses of coins by Mr J. A. Phillips :—

L

Roman As, B.C. 500—

Copper,	.	.	.	.	. 69·51
Tin, .	.	.	.	.	. 7·10
Lead, .	.	.	.	.	. 22·02
Iron, .	.	.	.	.	. ·48
Cobalt,	.	.	.	.	. ·59

99·70

Semis, B.C. 500—

Copper,	.	.	.	.	. 62·05
Tin, .	.	.	.	.	. 7·62
Lead, .	.	.	.	.	. 29·35
Cobalt,	.	.	.	.	. ·23
Nickel,	.	.	.	.	. ·19
Iron, .	.	.	.	.	. ·17

99·61

Quandaus, B.C. 500—

Copper,	.	.	.	.	. 72·17
Tin, .	.	.	.	.	. 7·17
Lead, .	.	.	.	.	· 19·52
Iron, .	.	.	.	.	. ·41
Cobalt,	.	.	.	.	. ·29
Nickel,	.	.	.	.	. ·20

99·76

Alexander the Great, B.C. 335—

Copper,	.	.	.	.	. 86·73
Tin, .	.	.	.	.	. 13·15
Sulphur,	.	.	.	.	. ·7

99·75

Julius and Augustus Cæsar—

Copper,	.	.	.	.	. 78·88
Tin, .	.	.	.	.	. 7·95
Lead, .	.	.	.	.	. 12·80

99·63

However, there are coins later than these that are purer alloy. For instance, an Attic coin analysed by Mitscherlitz, gave—

Copper,	.	.	.	.	. 88·46
Tin, .	.	.	.	.	. 10·04
Lead, .	.	.	.	.	. 1·05
					99·55

Coin of Alexander the Great, analysed by Schmid, gave—

Copper,	.	.	.	.	. 95·96
Tin, .	.	.	.	.	. 3·28
Lead, .	.	.	.	.	. ·76
					100·00

In the numerous analyses of coins we find no fixed proportions. Different kings seem to have made a change in the quality of their coins, according to the state of the national exchequer, and quantity or value of each metal. There is one thing worthy of notice in the above analyses, viz., that those of the same date 500 B.C., are very nearly the same quality, and they have all cobalt and nickel, while these metals do not appear in the coins 160 years later, made by a different nation. Analyses of coins might be multiplied to a considerable extent, but without adding much to the present inquiry.

Let us now direct our attention to the bronze articles mentioned in the Bible. In the description of the formation of the laver for the tabernacle, occur the following words—"And he (Bezaleel) made the laver of brass, and the foot of it of brass, of the looking-glasses of the women assembling, which assembled at the door of the tabernacle of the congregation " (Exod. xxxviii. 8).

The word translated *brass* should be *bronze*, that translated *foot* is rendered in a marginal note *cover*, and that translated *looking-glasses* is rendered in another note *mirrors*. As it was a bronze *foot* or *cover* which was manufactured from these mirrors, it is self-evident that the mirrors must have been metallic. And as we know that it was customary in very early times for women to wear as an article of dress a bronze mirror, we have no hesitation in concluding that these "looking-glasses of the women" were bronze mirrors.

Bronze suitable for mirrors must contain a large proportion of tin. Our speculum metal for telescopes, which yields the most perfect mirror yet known, consists, in round numbers, of two parts copper, and one part tin. An ancient Etruscan mirror, analysed by Gerard, gave—

Copper, ...	. 67·12
Tin, ...	. 24·93
Lead, ...	. 8·13
	100·18

Another ancient mirror, analysed by Klaproth, gave—

Copper,	.	.	.	.	.	62
Tin,	.	.	.	.	.	32
Lead,	.	.	.	.	.	6
						100

That the *cover* or lid—*foot*, we consider an incorrect translation—was made of a special quality of bronze, is significant of at least two things. 1st, It indicates that at this very early period special qualities of bronze were known and manufactured for distinct purposes. 2d, The particularising of mirror bronze gives the clue to a purpose beyond that of being merely a lid, which this cover was intended to serve, viz., that of a mirror. The laver was the basin or wash-stand in which the priests washed, and the cover on being turned back would serve as a mirror in which they could examine themselves before approaching the altar.

The altar of burnt-offering, or brazen altar as it is termed, was made of boards, and then overlaid with bronze. The size of this altar, taking Sir Isaac Newton's measure of a cubit, was ten feet five inches square, and six feet three inches high. The horns of this altar, and the staves or poles for carrying the fire grate, were also made of wood, overlaid with bronze. The bronze covering this large platform and stands, was no doubt cast into plates and hammered thin,—we say this advisedly, both from the necessity of the case, and also from the comparatively small quantity of bronze that was used; and it is indicative of skill in the manufacture of the alloy, for we have already shown that it is only bronze of certain qualities that have this property. All the articles that were to come into contact with the fire during sacrifice, were made of or covered with bronze. Of these we have the grate of network, fire-pans, shovels, tongs, pots, basons, and the rings attached to the grate, through which the staves were put for carrying it with the fire. There were also made of bronze flesh hooks, pins for staying the tabernacle, and what are termed the sockets into which were inserted the pillars or uprights, on which was suspended the roof and curtains of the outer court of the tabernacle, and the door. These were no doubt masses of bronze with socket holes for the pillars, which either stood upon, or were fixed into the ground. To cast a mass of metal with a socket hole, whether a round or square hole, would require a core—an operation that shows a considerable advance in the knowledge of casting. The discovery of casting in core is claimed by the Greeks, centuries after this date. The weight of these bronze sockets is not given, but those made for the pillars of the inner court, which were silver, were each one talent of silver, equal to 103 lbs. avoirdupois. Supposing the bronze sockets to be of the same weight, we find from Exod. xxxvi. 38, that there were five for the pillars of the tabernacle door, and in chap. xxxviii., we find mention of pillars with the sockets of bronze, these at 103

lbs. each give a gross weight of fifty-two cwts.
The whole weight of bronze used in the tabernacle
is given at 70 talents and 2400 shekels, equal to
7282 lbs. avoirdupois, or sixty-five cwts.; taking
fifty-two cwts. from this leaves 13 cwts. for tent
pins, furniture, laver, and overlaying of the altar;
which we think warrants our statement that the
sheets overlaying the wood must have been
thin. We have no indication that they used
rollers in these times for rolling out metal into
sheets, but they were certainly expert with the
hammer.

In the chapter on gold and silver, some idea of
the great amount of precious metals used in the
construction, and in the manufacture of the vessels
of the temple was given, but there is no statement
by which we can determine the quantity of bronze
used in this vast structure; either from the vastness
of the amount, or because it was of comparatively
little value. We find it stated, " And Solomon
left all the vessels unweighed, because they were
exceeding many, neither was the weight of the
brass (bronze) found out; " evidently from their
number, and probably from their size, for many
of the bronze productions were of immense weight.
It is impossible for any one to read over the
graphic account given in the Scriptures, of the
construction of the temple, especially of the productions in metal, without being struck, not only
with the number of the vessels and largeness of

the castings, but with the great variety of the work done.

First we may mention the two bronze pillars, the dimensions of which we give in English measure, taking the cubit to be according to Sir Isaac Newton's calculation, equal to twenty-five inches. In the account of the dimensions of these there is some discrepancy in 1 Kings vii. 15, 16. We read, " For he cast two pillars of brass, of eighteen cubits high a piece : and a line of twelve cubits did compass either of them about. And he made two chapiters of molten brass, to set upon the tops of the pillars : the height of the one chapiter was five cubits, and the height of the other chapiter was five cubits." Again in 2 Chron. iii. 15, we find the passage, " Also he made before the house two pillars of thirty and five cubits high, and the chapiter that was on the top of each of them was five cubits." Commentators have endeavoured to reconcile these discrepancies. The first is the measurement that is generally agreed upon, and which we adopt.

These pillars, then, without the capitals measured thirty-seven and a half, and in round numbers eight feet in diameter ; if they were hollow, as Whitson in his translation of Josephus thinks they were, the metal would not be less than three and a half inches thick, so that the whole casting of one pillar must have weighed from twenty to twenty-five tons. The height of the capitals was

ten feet five inches, which reckoning them also to
have been hollow, and of the same thickness as
the pillars, would give a weight of about ten tons
each. From the following description, all practical men will be able to appreciate the elaborate-
ness of the workmanship, in the finishing of these
capitals with their pillars. " And nets of checker
work, and wreaths of chain work, for the chapiters
which were upon the top of the pillars; seven for
the one chapiter, and seven for the other chapiter.
And he made the pillars, and two rows round
about upon the one network, to cover the chapiters
that were upon the top, with pomegranates: and
so did he for the other chapiter. And the
chapiters that were upon the top of the pillars
were of lily work in the porch, four cubits. And
the chapiters upon the two pillars had pomegranates also above, over against the belly which was
by the network: and the pomegranates were two
hundred in rows round about upon the other
chapiter. And upon the top of the pillars was
lily work! so was the work of the pillars finished."
The pillars when set up would measure about
forty-seven feet in height.

The next great work is the bronze altar, which
measured forty-one feet eight inches square, and
twenty feet ten inches high, forming a platform
of seventeen hundred and thirty-six square feet.

Some Scripture commentators are of opinion
that this altar was made in the same way as the

brazen altar in the tabernacle, of wood and overlaid with bronze.   We think that if this had
been the case, mention would have been made
of the fact, as mention is made when wood is
overlaid with gold, and are inclined to think
that the altar was entirely of bronze.   It is
thus described: "Moreover he made an altar of
brass, twenty cubits the length thereof, and twenty
cubits the breadth thereof, and ten cubits the
height thereof."   The thickness of the metal used
for this altar is nowhere given, but supposing it to
have been three inches, the whole weight of the
metal would not be under two hundred and fifty
tons.

The next large casting is that termed the
molten sea: "Also he made a molten sea of ten
cubits from brim to brim, round in compass, and
five cubits the height thereof; and a line of thirty
cubits did compass it round about.   And under it
was the similitude of oxen, which did compass it
round about: ten in a cubit, compassing the sea
round about.   Two rows of oxen were cast when
it was cast.   And the thickness of it was an hand-
breadth, and the brim of it like the work of the
brim of a cup, with flowers of lilies; and it received and held three thousand baths."   This immense hemispheric vessel would therefore measure
twenty-one feet eight inches in diameter, and be
ten feet five inches deep in the centre.   This large
casting could not weigh less than thirty tons, and

would be capable of holding twenty thousand gallons of water. The brim had cut out upon it, or was ornamented with "flowers of lilies," probably cast and fixed upon the brim as if growing, and oxen were carved or cut on the outside all round, to the number of 300, and it stood upon a pedestal of twelve bronze oxen:—"It stood upon twelve oxen, three looking toward the north, and three looking toward the west, and three looking toward the south, and three looking toward the east: and the sea was set above upon them, and all their hinder parts were inward."

The size of these oxen or bulls is not given, but they must have been of considerable size, so as that their corresponding legs would give thickness and strength, sufficient to support so great a weight; for when the vessel was filled with water, the whole weight would be upwards of one hundred tons. How appropriate would such legs as those described by Dr Percy and found by Mr Layard be, where iron was found overlaid with bronze; however, bronze can bear an enormous weight before crushing.

There were also castings of a smaller sort, the details of which, as given by the sacred narrator, show them to have been castings of very considerable size, as for instance:—"And he made ten bases of brass; four cubits was the length of one base, and four cubits the breadth thereof, and three cubits the height of it. The work of the

bases was on this manner : they had borders, and
the borders were between the ledges : and on the
borders that were between the ledges were lions,
and oxen, and cherubims : and upon the ledges
there was a base above : and beneath the lions
and oxen were certain additions made of thin
work.   And every base had four brazen wheels
and plates of brass ; and the four corners thereof
had undersetters : under the laver were undersetters molten, at the side of every addition.   And
the mouth of it within the chapiter and above was
a cubit : but the mouth thereof was round after
the work of the base, a cubit and a half : and also
upon the mouth of it were gravings with their
borders, foursquare, not round.   And under the
borders were four wheels ; and the axle-trees of
the wheels were joined to the base : and the height
of a wheel was a cubit and half a cubit.   And
the work of the wheels was like the work of a
chariot wheel : their axle-trees, and their naves,
and their felloes, and their spokes, were all molten.
And there were four undersetters to the four
corners of one base : and the undersetters were of
the very base itself.   And in the top of the base
was there a round compass of half a cubit high :
and on the top of the base, the ledges thereof and
the borders thereof were of the same.   For on the
plates of the ledges thereof, and on the borders
thereof, he graved cherubims, lions, and palm trees,
according to the proportion of every one, and

additions round about. After this manner he made the ten bases : all of them had one casting, one measure, and one size " (1 Kings vii. 27–37).

These bases, fitted upon wheels, constituted carriages for supporting and moving about the ten lavers, or large bronze vessels for washing, &c. The whole workmanship of these bases, as described, is indicative of great skill. The lavers are thus described : "Then made he ten lavers of brass ; one laver contained forty baths, and every laver was four cubits, and upon every one of the ten bases was one laver." When it is considered that each of these vessels were capable of holding three hundred gallons of water, upwards of a ton weight of water each, we obtain a better idea of their size. Each vessel, upon its carriage, and full of water, would weigh no less than two tons. We need not specify more of these large works constructed for the temple ; those named are sufficient to verify what has been stated in reference to the knowledge of working in bronze possessed by these artisans. A due estimate of the massive character of the castings made, must lessen the surprise felt on reading the statement, that the weight of the brass was not known from its abundance. That different qualities of bronze were made and used for different purposes, for beauty or suitability, is apparent from the following general passage (1 Kings vii. 45), " And the pots, and the shovels, and the basins : and all these vessels, which Hiram made to king Solomon for the house of the Lord, were of bright brass." These were in all probability made of bronze, containing a larger proportion of tin; an alloy of the character of speculum metal, which when polished would present an imposing appearance.

Any practical metallurgist who reads carefully the description given of the metallic work, executed for the temple, must form no mean estimate of the skill of these ancient workers in the metallurgical arts, not only in casting, but in the delicate and intricate operations of engraving, carving, and chasing. It must also be apparent, in reading over the account of these enormous castings, that vast mechanical resources must have been within their reach, in order to remove, fit up, and place in position, such vast masses of metal.

Two questions arise: Where were these castings made, and of what material were the moulds? In reply to these questions, we find the following statement (1 Kings vii. 46):—

" In the plain of Jordan did the king cast them, in the clay ground between Succoth and Zartham." According to the marginal reading, the translation should be: " In the depth of the clay ground," showing that they were moulded in clay. Some have suggested that the smelting and casting operations were performed outside Jerusalem for sanatory purposes. We scarcely think the smoke question troubled the dwellers in Jerusalem, and are more inclined to accept the reasons, which the statement of the sacred historian naturally suggests. Clay mixed with sand is the moulding materials still used for bronze castings. So large a quantity of metal as some of these castings required would not, in all probability, be fused in one furnace. For all the large castings, especially for such a massive casting as the " sea of brass," it is highly probable that a whole series of furnaces were put in operation at the same time, and all tapped together, and the molten metal run into one mould. Such series of furnaces are generally set in a sort of circle or square, under a large dome or roof, from which rises a large chimney or tower. These structures are generally in the neighbourhood of towns, and always present a very conspicuous appearance. Now, if a similar method was adopted in these early days, it may explain the reference of Nehemiah, in the following passage :—" Malchijah the son of Harim, and Hashub the son of Pahath-moab, repaired the other piece, and the tower of the furnaces." This may imply that such a structure as we have just described did at one time exist on the plains of Jordan, erected probably in the time of Solomon, and it may have continued a place for metallurgical operations—a national foundry—up to the time the Jews were carried captive into Babylon. Again, at the restoration, or rather during the rebuilding and refurnishing of the temple, the same metallurgical operations would be resumed, hence the repairing of these furnaces would form an important object to the returned captives ; and as they came out from the midst of a people who were well skilled in metallurgical operations, as is evidenced by such work as the golden image in the plains of Dura, and the researches of Layard and others, it is probable that many of the Jews were well skilled in the metallic arts. When we consider the enormous quantity of bronze, which must have been used in the construction and furnishing of the temple, and that about ten per cent. at least of the whole must have been tin, we are led to form some conception of the abundance of this metal at that period. To us this matter has a home interest from the circumstance, that it is all but certain that the tin used for the work of the temple was found in our own island. As tin is not named separately, it is probable that the Phœnicians made and sold the bronze, and that it was not manufactured by the Jews ; and this is made more probable from the absence of articles made wholly of tin being found in ancient ruins.

The knowledge of metallic arts possessed by the Israelites when they left Egypt, but which during the time of the Judges had, on account of the special circumstances of the nation, been allowed to lapse till such knowledge was practically lost, was again recovered during the building of the temple, being then introduced by the Phœnicians.

After this time some knowledge of metallic arts appears to have been retained by them, and never afterwards lost; for after this period we find frequent references in Scripture to smiths and other artificers in metals, and also we find mention made of particular metallic works requiring considerable skill, which were manufactured by themselves. With the exception, however, of these two periods—viz., the construction of the tabernacle and the temple—the Hebrews do not appear to have ever been a manufacturing people, or to have excelled in any metallic arts.

There were certain ancient temples and palaces which contained immense wealth in the quantity of metals with which they were furnished and adorned, but we have not seen an account of any single building on which there was such an expenditure of metal of all kinds as the temple of Solomon. Such vast treasures, accumulated in one building, were well calculated to excite, and did excite, the cupidity not only of wicked and extravagant rulers, but of the monarchs of other nations. It will be interesting to ascertain, if possible, what became of such a vast quantity of metal thus brought together into the house of the Lord.

The temple was commenced in or about the year 1011 B.C., and forty-one years after this date, 970 B.C., Shishak, king of Egypt, captured Jerusalem, and plundered the temple and palace of much gold, taking away the shields and targets

M

of gold, which were afterwards replaced by
targets and shields of bronze, an account of
which is given in 1 Kings xiv.—"And it came
to pass, in the fifth year of King Rehoboam,
that Shishak, king of Egypt, came up against
Jerusalem : and he took away the treasures of the
house of the Lord, and the treasures of the king's
house; he even took away all : and he took away all
the shields of gold which Solomon had made. And
King Rehoboam made in their stead brazen shields,
and committed them unto the hands of the chief of
the guard which kept the door of the king's house."

The extent of this spoliation, we think, requires a
little modification. The reading of the passage is
clearly that the King of Egypt took away all the treasures of "the house of the Lord, and the treasures
of the king's house; " but the word treasure has
evidently a limited meaning. It probably means all
the wealth that was stored up as bars of gold and
silver. The only manufactured articles named
which he seems to have taken were the targets and
shields. The value of the targets and shields would
be somewhere about £230,000. It is said the
rulers of Israel humbled themselves, and became
servants or tributaries to Shishak; so that probably
the sums or articles he got were matters of agree-
ment, or it required all the metals stored up
with the shields and targets to pay the tribute
demanded.

In the meantime Judah had been led into
idolatry by Abijah and his wife, the mother of

King Asa. Asa, when he became king, removed his mother from being queen, and put down her idolatries, destroyed her idol, stamped it and burned it, and turned his heart to the worship of God. It is said (2 Chron. xv. 18), "He brought into the house of God the things that his father had dedicated, and that he himself had dedicated, silver and gold and vessels." Asa reigned peaceably until 920 B.C.; when he was besieged by the King of Israel. In this strait he made free use of the treasures of the temple to bribe the King of Syria to assist him against Israel. "And Baasha, king of Israel, went up against Judah, and built Ramah, that he might not suffer any to go out or come in to Asa, king of Judah. Then Asa took all the silver and the gold that were left in the treasures of the house of the Lord, and the treasures of the king's house, and delivered them into the hand of his servants: and King Asa sent them to Ben-hadad the son of Tabrimon, the son of Hezion, king of Syria, that dwelt at Damascus, saying, There is a league between me and thee, and between my father and thy father: behold, I have sent unto thee a present of silver and gold; come and break thy league with Baasha, king of Israel, that he may depart from me" (1 Kings xv. 17–19). This present of gold and silver evidently did not embrace any of the works of art, but only that which was stored in the treasury of the temple.

Jehoshaphat, the son of Asa, seems to have

turned his attention to raising his kingdom in
arts and commerce: he built new cities, and
fortified those his father had built; he also built
forts, and established storehouses throughout the
land; he appointed learned men to teach the
people the law, which embraced moral, civil, and
religious duties, for the purpose of re-establish-
ing that commerce in gold which flourished so
greatly in the days of Solomon; he entered into
an alliance with Ahab, the king of Israel, who was
married to a princess of Phœnicia, the then great
maritime nation of the world, and at Ahab's
death he renewed this alliance with Ahaziah,
Ahab's son. In order to further this project, he
built large ships that were intended to sail to
Ophir; but we suspect Ahaziah was not so true
and faithful an ally as Hiram was to Solomon,
for these vessels never went out on their voyage,
but were wrecked in the harbour where they were
built, of Eziongeber, on the Red Sea. The union
of Joram, Jehoshaphat's son, with the daughter
of Ahab, Athaliah, a determined idolatress, who
ruled the nation through her husband, and at
his death murdered the seed royal and had
herself proclaimed queen, was a source of much
evil to Judah. During her reign the temple
was allowed to get out of repair, while she
used much of its treasures for the service of
her idolatrous worship. From this neglect and
spoliation, the temple, or certain portions of it,
suffered great injury. "For," says the record,

"the sons of Athaliah, that wicked woman, had broken up the house of God, and also all the dedicated things of the house of the Lord did they bestow upon Baalim." The disrepair and spoliation of the temple at this time seems to have been so great as to unfit it for the proper worship of God, as prescribed by the law; for at the death of Athaliah, which took place about 870 B.C., it was found necessary to have it repaired, and for this purpose a voluntary contribution or collection was made, the contributions being dropped into a box or chest placed in the temple, as is stated in 2 Chron. xxiv. 11-14 — "And thus they did day by day, and gathered money in abundance. And the king and Jehoiada gave it to such as did the work of the service of the house of the Lord, and hired masons and carpenters to repair the house of the Lord, and also such as wrought in iron and brass to mend the house of the Lord. So the workmen wrought, and the work was perfected by them, and they set the house of God in his state, and strengthened it. And when they had finished it, they brought the rest of the money before the king and Jehoiada, whereof were made vessels for the house of the Lord, even vessels to minister, and to offer withal, and spoons, and vessels of gold and silver."

When King Ahaz visited the King of Assyria, he saw at Damascus an altar of superior workmanship that pleased him so much, that he sent a pattern or drawing of it to Jerusalem, with orders

to have one made for him exactly similar against
his return, that he might worship thereon—" And
King Ahaz went to Damascus to meet Tiglath-
pileser, king of Assyria, and saw an altar that was
at Damascus : and King Ahaz sent to Urijah the
priest the fashion of the altar, and the pattern
of it, according to all the workmanship thereof.
And Urijah the priest built an altar according to
all that King Ahaz had sent from Damascus : so
Urijah the priest made it against King Ahaz came
from Damascus. And when the king was come
from Damascus, the king saw the altar : and the
king approached to the altar, and offered thereon.
And he brought also the brazen altar, which was
before the Lord, from the forefront of the house,
from between the altar and the house of the Lord,
and put it on the north side of the altar." This was
followed immediately by a sacrilegious act of the
grossest kind—" And King Ahaz cut off the
borders of the bases, and removed the laver from off
them, and took down the sea from off the brazen
oxen that were under it, and put it upon a pavement of stones " (2 Kings xvi.)

Dr Kitto, in his " History of Palestine," thinks
that these bases, and also the twelve oxen, were
probably melted down for idolatrous purposes.
The twelve oxen, however, were not melted, for they
are named by Jeremiah among the articles taken
away to Babylon. Probably the oxen were used
for some idolatrous object.

On this subject the Rev. Mr Rawlinson, in his

"Ancient Monarchies," says that "Tiglath-pileser was a very religious person, which was manifested in the number and great gifts he conferred upon his gods. And according to the usual mode of reasoning adopted, his success and great prosperity made it evident that not only his earnestness in his devotions, but the gods he worshipped were really the true gods. And this would evidently sway mightily with a weak-minded monarch such as Ahaz. As bronze bulls and oxen were held in high esteem in Babylon, symbolising their worship, it is more than probable that the oxen were preserved. Indeed it is very probable that the brazen laver was removed for the purpose of obtaining the oxen on which it stood."

We are of opinion, from reading the statement in 2 Kings xvi. 17, that along with the borders of the bases cut off, the metal of the large bronze altar was used up by Ahaz for altars or idols about the city of Jerusalem, for the temple was shut up—"And Ahaz gathered together the vessels of the house of God, and cut in pieces the vessels of the house of God, and shut up the doors of the house of the Lord, and he made him altars in every corner of Jerusalem."

In the catalogue given by Jeremiah of the articles taken away to Babylon, the brazen altar is not named; this we think would not have been omitted if it was then in existence.

Hezekiah succeeded Ahaz, and in his reign the temple was again opened and cleaned, and the

vessels which Ahaz had cast aside and polluted
were purified—"Moreover all the vessels which
King Ahaz in his reign did cast away in his
transgression have we prepared and sanctified;
and behold, they are before the altar of the
Lord." Also, "He (Hezekiah) removed the high
places, and brake the images, and cut down the
groves, and brake in pieces the brazen serpent
that Moses had made; for unto those days the
children did burn incense to it: and he called it
Nehushtan." Nehushtan is stated in the 'margin
to mean a *"piece of brass."* Possibly Hezekiah
had the brazen sea replaced upon the oxen, as it
was before.

We find mention made here of the *altar of the
Lord.* Some have supposed that this was the original large bronze altar; but "altar of the Lord"
would be the name given to whatever altar existed
in the temple at the time, and does not necessarily
signify the large altar removed by Ahaz.

Even good King Hezekiah despoiled the temple
of part of its fittings; but this was in an emergency, and for the safety of the nation. His non-
payment of the stipulated tribute-money to the
King of Assyria provoked the anger of that
powerful monarch, who, in the fourteenth year of
Hezekiah's reign, poured his troops into Judea,
and took the fenced cities. At this Hezekiah's
courage failed him, and he sent to the King of
Assyria this humiliating message — "I have
offended: return from me; that which thou put-

test on me I will bear." Sennacherib exacted
three hundred talents of silver and thirty talents
of gold. Taken as Jewish talents, this would be
about £266,906. We think that this sum was
the amount of the tribute-money which was due,
but that the Assyrians also exacted a large sum
to remunerate them for withdrawing from the
campaign; for it is said that "Hezekiah gave him
all the silver that was found in the house of the
Lord and in the treasures of the king's house;
and he cut off the gold from the doors of the
temple of the Lord, and from the pillars which
Hezekiah king of Judah had overlaid, and gave
it to the King of Assyria." We can scarcely
imagine that the exchequer of the nation was so
poor at this time that they could not pay so
comparatively small a sum—£266,906—without
despoiling the temple. He probably adopted this
method to meet the present emergency until
a call was made upon the nation, which supposition is strengthened by the fact that Hezekiah
very quickly recovered from this loss of wealth;
for we find it stated shortly afterwards, "And
Hezekiah had exceeding much riches," of which
he made a boastful show to the ambassadors of
the King of Babylon.

Nearly a century after this exhibition of wealth,
Nebuchadnezzar, after a protracted siege, reduced
Jerusalem, destroyed the temple, and carried away
most of the metals to Babylon—" And the pillars
of brass that were in the house of the Lord, and

the bases, and the brazen sea that was in the
house of the Lord, did the Chaldeans break in
pieces, and carried the brass of them to Babylon.
And the pots, and the shovels, and the snuffers,
and the spoons, and all the vessels of brass wherewith they ministered, took they away. And the
firepans, and the bowls, and such things as were
of gold in gold, and of silver in silver, the
captain of the guard took away. The two pillars,
one sea, and the bases which Solomon had made
for the house of the Lord; the brass of all these
vessels was without weight." Jeremiah, furnishing
a catalogue of the same articles, says—"And the
basins, and the firepans, and the bowls, and the
caldrons, and the candlesticks, and the spoons, and
the cups; that which was of gold in gold, and that
which was of silver in silver, took the captain of the
guard away. The two pillars, one sea, and twelve
brazen bulls that were under the bases, which
King Solomon had made in the house of the Lord:
the brass of all these vessels was without weight."
The large bronze altar is not here mentioned.
Whether the gold used in overlaying the house,
and in making up the furniture of the temple,
was all removed previous to this, is not certain;
but from what is stated, it would appear that
much gold remained. The removal of such a vast
quantity of metal to Babylon must have been a
formidable undertaking.

Many of the sacred vessels of gold and silver
were preserved from destruction, and carried to

Babylon by the Assyrians, who placed them in
.the temple of their idols. Very shortly after the
removal of the gold, silver, and brass from the
temple in Jerusalem, "Nebuchadnezzar the king
made an image of gold, whose height was threescore cubits, and the breadth thereof six cubits:
he set it up in the plain of Dura, in the province
of Babylon."

This taking place so shortly after the fall of
Jerusalem suggests the supposition that the image
may have been made from the metal removed
from that city. The siege ·had been a formidable
undertaking, and sufficiently important to warrant
a memorial being erected; and the image, besides
being set up in honour of some god, was in all
probability not only for the purpose of being worshipped, but might at the same time commemorate
the defeat of the `Jews and the taking of Jeru-
salem. This suggests a reason why the Jews were
watched during its dedication, and why the three
young patriots did not go to the dedication of
the image—attendance being not only an acknow-
ledgment of the superiority of the Assyrian gods
over the God of Israel, but a rejoicing at their
own defeat. It is quite within the range of probability that some of the bronze articles found by
Mr Layard and others in the ruins of Babylon
may have been either part of the furniture of, or
made from the bronze formerly in, the temple of
Solomon.

In the account of this sack of the temple, it is

remarkable that no mention is made of the Ark of the Covenant, no word to say whether it was carried to Babylon or destroyed. In either case, it is more than probable that such an event would have been referred to by Jewish historians, as the great national reverence for this object would have induced them to have lamented either its destruction or removal. We are inclined to think that this sacred vessel was hid by pious priests, who, foreseeing certain destruction coming upon the city and temple, removed it to a place of safety, and their death following, the locality of its hiding-place was forgotten.

When the Jews returned from Babylon to their own land, B.C. 536, they obtained permission to carry back with them the sacred vessels of their temple which had been preserved in the temples of Babylon since their captivity. It is thus narrated by the historian of that event—" Also Cyrus the king brought forth the vessels of the house of the Lord which Nebuchadnezzar had brought forth out of Jerusalem, and had put them in the house of his gods; even those did Cyrus, king of Persia, bring forth by the hand of Mithredath the treasurer, and numbered them unto Sheshbazzar, the prince of Judah. And this is the number of them :—Thirty chargers of gold, a thousand chargers of silver, nine and twenty knives, thirty basins of gold, silver basins of a second sort four hundred and ten, and other vessels a thousand. All the vessels of gold and of silver were five

thousand and four hundred. All these did Sheshbazzar bring up with them of the captivity that
were brought up from Babylon to Jerusalem" (Ezra
i. 7–11). These were simply what were brought back
from Babylon; but the Jews, with an enthusiasm
that always characterised them in relation to their
temple, gave liberally of their substance. Although
at this time they were doubtless much impoverished,
the various gifts in gold and silver, and vessels
made of the same recorded in Ezra and Nehemiah,
amounted to close on two millions of our money,
exclusive of garments, and other articles in metal
not valued. And they continued to enrich their
temple with the precious metals until its final
destruction. The most of what may be termed
the furniture or furnishing of the second temple
remained in it until it was destroyed by Titus after
the Christian era. These were then carried in
triumph to Rome, as Josephus, who was an eye-
witness, testifies—" But for those that were
taken in the temple of Jerusalem, they made the
greatest figure of them all. The golden table, of
the weight of many talents; the candlestick that
was made of gold, though its construction was now
changed from that which we made use of—for its
middle shaft was fixed upon a bases, and the small
branches were produced out of it to a great length,
having the likeness of a trident in their position,
and had every one a socket made of brass for a lamp
at the top of them. These lamps were in number
seven, and represented the dignity of the number

seven amongst the Jews " (Josephus' " War with the Jews," Book vii., Whitson's translation).

The figure of the candlestick was sculptured on the triumphal arch of Titus, which is still to be seen.   It appears, however, from another extract from Josephus, that there were more candlesticks in the temple than one.   He says that " a priest named Jesus, a son of Thebuthus, on condition of having his life spared, brought out to Titus from the sacred treasury two candlesticks formed like the candlestick of the temple, some tables, cups, and other vessels, all of solid gold, and very heavy, as also the sacred veils, the official robes of the high priest, ornamented with precious stones, and many of the sacred utensils.   Phineus, the treasurer of the temple, who also was taken prisoner at this time, delivered to Titus the robes and girdles of the priests, a great quantity of purple and scarlet, which was preserved for the veils, and also cinnamon, cassia, and other sweet spices, which were used for incense."

We are inclined to think that neither the candlestick carried by Titus, nor any of them in the temple at the time of its destruction, was the candlestick made in the wilderness.   That this was that candlestick, however, has often been assumed : the historical evidence is against that assumption.   There is no mention of this candlestick having been carried to Babylon, nor does it appear in the list of the articles brought back ; such an important item, we think, would not have been omitted in both lists. The same remark applies to the table. We think that the candle-stick of Moses was removed from the temple before its destruction by sacrilegious or friendly hands. The candlestick represented on the arch of Titus does not correspond with that made by Moses, and the table is neither in shape nor dimensions that made in the wilderness. It may represent the altar of incense.

After the triumph, the candlestick taken by Titus was deposited in the Temple of Peace, and according to one story, fell into the Tiber from the Milvian bridge, during the flight of Mexentius from Constantinople, in 312 A.D. Gibbon says that Genseric sacked Rome in 455 A.D., and took the candlestick with him to Carthage; and Belisarius finding it there when he defeated the Vandals, brought it to Constantinople. Then in 534 A.D. it was deposited, with other vessels of the Jewish temple, in the Christian Church at Jerusalem. It has never been heard of since, but it is possible it may be found yet, when we get free access to the sacred places to excavate; and if found, it would certainly form one of the most interesting relics of antiquity, even although not that of the taber-nacle. The golden tables referred to by Josephus no doubt found their way into the national exchequer of Rome, to help them to carry on those wars which ultimately brought about the ruin of the Roman Empire itself.

IRON.

Iron is the most universally diffused of the metals, being present in some of its combinations in almost every substance in nature, whether animal, vegetable, or mineral. It is very seldom found in the metallic state. Its most important combinations, in a metallurgic point of view, are with oxygen and carbonic acid, and in these combinations it is found in sufficient abundance for all our wants. The reduction of these ores to the metallic state is performed by the same means as described for the oxides of other metals—namely, by mixing with coal, and other carbonaceous matters, and subjecting to a heat of sufficient intensity to cause fusion. The greater portion of the iron made in this country at the present time is obtained from the carbonate. This carbonate is found mixed or combined either with coaly matter, in which case it is termed blackband, or with clayey matters, which are termed clay or slate band. In either case the ore is broken into small pieces; and if there is not sufficient coaly matter in the ore, it is mixed with coal, so as to burn. The whole is made into a large heap, and set on

fire, which dispels carbonic acid as well as carbon,
and leaves the iron as oxide, with less or more
earthy matters, as clay and silica, which are
got rid of by mixing with the ore a quantity of
lime, which during the process of fusion combines
with the earths, and forms scoria or slag. This
floats on the surface of the fused mass, and is
separated from the iron by being let out of the
furnace at a different part. The iron thus obtained
is known as cast iron, and is afterwards, when
required, converted into malleable iron by repeated heating and hammering. To effect the
fusion of cast iron, we must have furnaces of very
large dimensions, and powerful steam-engines for
blowing the fire, before the heat is rendered
sufficiently intense to reduce the ore. And here
lies the interest in connection with the inquiry
into the metallurgical operations of iron in ancient
times, back even to the infancy of the human
race. The smelting and manufacture of iron by
this method is surrounded with so many difficulties,
and needs so many requirements and such skill,
that were there no other means of obtaining it than
by this one process, of first making cast iron and
converting that into malleable, we would naturally
expect it to have been among the last of the
metals brought into use. And it is no doubt this
view of the question that has caused so much
scepticism concerning the early use of iron. Instead
of it being the last of the metals brought into use,

we find it among the first named in history—" And
Zillah, she also bare Tubal-cain, an instructor of
every artificer in brass and iron " (Gen. iv. 22).
In Thomson's " History of Chemistry," the Doctor,
knowing the difficulty connected with this very
early knowledge of iron, suggested " that in
these early days it is possible native iron may have
existed, as well as native gold, silver, and copper,
and in this way Tubal-cain may have become
acquainted with its existence and properties."
But as it must have been a more difficult task
for Tubal-cain to make bronze than iron, we are
afraid that the Doctor has chosen his standpoint
from his knowledge of our requirements, and has
forgotten that oxide of iron can be easily reduced to
the metallic state although not fused, even at an
ordinary red heat, and the metal made fit for use,
by hammering also at a red heat.   This matter we
shall have occasion to speak of further on.   We
think, therefore, the Doctor's supposition that
Tubal-cain had native iron to work upon is not a
very happy one, seeing there is scarcely an instance
of iron being found native, and then only in quantities so small as to be of no practical value.   Even
allowing it to have been more abundant in the
early ages of man's history, the difficulties of free-
ing it from the matrix must have been about as
great as those in reducing it from the ore, unless
we allow a knowledge of the magnet, by which it
could be gathered from the rocks after they were

ground; but this would be an admission equally baseless, since the soil is so destructive to iron as to oxidise articles made of it in the course of a few centuries even in dry climates, so that all traces of them are lost, and this same destructive process would operate on native iron.

The method of manufacturing iron from its ore adopted by the Indians is very simple, and of a very ancient date, and probably is the same as was practised by the ancient Egyptians, and other nations.

The furnace is built as follows :—First a trench is made in the ground about three feet deep, having a sloping side or entrance, and then a furnace of brick or stone is made upon the side of the trench; the bottom of the furnace has an incline into the trench, and holes by which to remove the slag from the furnace. After a fire is kindled, the ore of iron, which is an oxide, is mixed with charcoal, and put into the fire; two bellows made of skins are connected to a clay-pot which projects from the bottom of the furnace, and is placed upon a plank, or planks, laid over the trench in front of the fire. A man sits and works the bellows by hand, pressing down one with the right hand, and then the other with the left, which produces a continuous stream of wind. In a short time the ore softens, a part of it melts along with the impurities, and forms slag or scoria, which is taken out by one of the holes or openings of the furnace. The remainder of the iron loses its oxygen by the charcoal, and forms a tough, pasty mass. About twelve hours after, the iron is withdrawn in blocks. The blocks of crude metal are then put into another furnace, and heated to a welding heat; it is then taken out and subjected to beating and hammering. This beating drives out any scoriaceous matters, and the metal toughens and refines. This method, with slight modifications in the shape of the furnace, arising probably from local circumstances, has been in use from the earliest ages, not only for iron, but for that compound of it termed steel, which will be apparent from the following interesting account of the manufacture of iron and steel in India given by Dr Ure :—

" The manner in which iron ore is smelted and converted into wootz, or Indian steel, by the natives at the present day is probably the very same that was practised by them at the time of the invasion of Alexander; and it is a uniform process from the Himalaya mountains to Cape Comorin. The furnace or bloomery in which the ore is smelted is from four to five feet high, somewhat pear-shaped, being about five feet wide at bottom, and one foot at top. It is built entirely of clay, so that a couple of men may finish its erection in a few hours, and have it ready for use the next day. There is an opening in front about a foot or more in height, which is built up with clay at the commencement, and broken down at

the end of each smelting operation. The bellows
are usually made of a goat's skin, which has been
stripped from the animal without ripping open the
part covering the belly. The apertures at the legs
are tied up, and a nozzle of bamboo is fastened
into the opening formed by the neck. The orifice
of the tail is enlarged and distended by two slips
of bamboo; these are grasped in the hand, and
kept close together in making the stroke for the
blast; in the returning stroke, they are separated
to admit the air. By working a bellows of this
kind with each hand, making alternate strokes, a
tolerably uniform blast is produced. The bamboo
nozzles of the bellows are inserted into tubes of
clay, which pass into the furnace—at the bottom
comes off the temporary wall in front. The furnace is filled with charcoal, and a lighted coal
being introduced before the nozzles, the mass in
the interior is soon kindled. As soon as this is
accomplished, a small portion of the ore, previously
moistened with water, to prevent it from running
through the charcoal, but without any flux whatever, is laid on the top of the coals, and covered
with charcoal, to fill up the furnace. In this
manner ore and fuel are supplied, and the bellows
are urged for three or four hours. When the process is stopped, and the temporary wall in front
broken down, the bloom is removed with a pair of
tongs from the bottom of the furnace.

" In converting the iron into steel, the natives

cut it into pieces to enable it to pack better in the
crucible, which is formed of refractory clay, mixed
with a large quantity of charred husk of rice.   It
is seldom charged with more than a pound of iron,
which is put in with a proper weight of dried
wood, chopped small, and both are covered with
one or two green leaves, the proportions being, in
general, ten parts of iron to one of wood and
leaves.   The mouth of the crucible is then stopped
with a handful of tempered clay, rammed in very
closely to exclude the air.   As soon as the clay-
plugs of the crucible are dry, from twenty to
twenty-four of them are built in the form of an
arch in a small blast-furnace; they are kept
covered with charcoal, and subjected to heat, urged
by a blast, for about two hours and a half, when
the process is considered to be complete.   The
crucibles being now taken out of the furnace, and
allowed to cool, are broken, and the steel is found
in the form of a cake, rounded by the bottom of the
crucible."   The remnants of some of these bole furnaces or bloomeries still exist in our own country.

These modes of manufacturing iron and steel
are, in the opinion of all who have studied the
subject, the same as those practised in the earliest
ages of the world; and although the processes are
simple, and the apparatus rude, still they are very
effective, and produce metals of the best kind.
Rude as the whole appears to be, yet it must have
been the result of considerable experience.   The

formation of the bellows—the composition of the crucibles—the charging of the furnaces and crucibles—all tell of experience and observation. There is no reference in Scripture or other ancient book to the methods practised by the ancients for obtaining iron from its ores; but that it was obtained and in common use in the very earliest ages is clearly proven by the constant reference to this metal and its properties in the oldest histories extant, and long before the age of any historian. The Scriptures are full of such references, showing that the Hebrews had knowledge of these practical arts. A few centuries after the Flood, taking the early date of Job, iron ore is referred to in the following language, " Iron is taken out of the earth." Moses refers to iron furnaces in terms which show that they have been in use, and well understood by those he was addressing. He makes the furnace the object of an allegorical figure expressive of intense suffering— " But the Lord hath taken you, and brought you forth out of the iron furnace " (Deut. iv. 20). The same writer also refers to the ore of iron as existing in Palestine—" Out of whose hills thou mayest dig iron and brass " (Deut. viii. 9). That it was a common article in different countries is evident from its being mentioned amongst articles taken from the Midianites, and concerning which general rules were given for purifying. In corroboration of these statements, and in order to

prove the early use of iron, and even of steel, the
following extract is taken from Wilkinson's
" Ancient Egypt: "—" Iron and copper mines are
found in the Egyptian desert which were worked
in old times ; and the monument of Thebes, and
even the tombs about Memphis, dating more than
4000 years ago, represent butchers sharpening
their knives on a round bar of metal attached to
their aprons, which, from its blue colour, can only
be steel.   And the distinction between the bronze
and iron weapons in the tomb of Rameses III.,
one painted red and the other blue, leaves no doubt
of both having been used at the same periods."

These facts harmonise with the reference in
Scripture to engraving on precious stones and
other works, which operation, so far as our knowledge goes, necessitates the use of both iron and
steel.   The fact of there being no iron articles
found in the ancient ruins of Egypt and Nineveh,
is easily accounted for by the rapid destruction of
that metal when exposed to air and moisture.
The iron instruments of war and art used by our
own forefathers a few centuries ago, that have
been subjected to such action, are now, when
found, mere fragments, with scarcely a trace of
their original shape, which a visit to any of our
public museums will verify ; so that the absence of
similar articles in very ancient ruins is easily
accounted for.   It is true that profane history says
little or nothing of iron in ancient times—a

circumstance often referred to in proof of iron not being known, or, if known, only obtained in very minute quantities. It is well known that the Scriptures, equally valuable and ancient, are seldom referred to when this subject is considered. Even many of those who profess their belief in the divine authenticity of the Scriptures, yet pass over such incidental references with indifference, and, to avoid noticing seeming difficulties, admit certain inferences upon art without much inquiry. In our opinion, numerous incidental references to arts and manufactures, or the products of these, are of very great importance, as illustrations of great truths in history, and also of the existence of arts which, because of an impression that these ages were barbarous, some cannot admit to have existed. Since in Scripture we find iron articles named as being in use for the ordinary purposes of life, and these the same purposes to which they are now put by ourselves, it is much stronger evidence, because positive, that iron was abundant, than is any negative evidence derived from the mere absence of mention of iron in the meagre history of other nations, or the non-existence of iron in ancient ruins. It is no stretch of fancy or of reason to conclude from such references that the method of extracting iron from its ore was well known, and so easily performed as to be no barrier to the common and everyday use of iron. If for its manufacture we require peculiarly-built

furnaces and strong blasts, the ancients must
have had means of producing the same effects;
but whether their appliances for this purpose were
the same as ours—whether their furnaces were
built of brick or stone, or their blast driven by
bellows or engine—or whether they used wood or
coal as fuel, or made one ton per day or per week
—we know not. The fact, however, is undeniable,
that they were in possession of means of overcoming many of the same difficulties that we overcome;
and that the advancement we have made is more
in the means of facilitating the process than in
any discovery of the reduction. But the extent of
their appliances is made somewhat apparent by
consideration of the several objects to which iron
was applied, as recorded in Scripture.

In warfare the use of iron weapons of various
kinds is named at a very early period, and is referred
to throughout the whole history of the Jewish
people, as the following quotations will show :—

"And all the Canaanites who dwell in the
valley have chariots of iron, both they who are of
Beth-shean and its towns, and they who are of the
valley of Jezreel" (Josh. xvii. 16). "And Judah
could not drive out the inhabitants of the valley,
because they had chariots of iron " (Judges i. 19).
"Jabin, king of Canaan, who reigned in Hazor,
the captain of whose host was Sisera, who dwelt in
Harosheth of the Gentiles. And the children of
Israel cried unto the Lord; for he had nine

hundred chariots of iron; and twenty years he mightily oppressed the children of Israel " (Judges iv. 2, 3). Even allowing, as some suppose, that the designation iron chariots means chariots with iron weapons, like scythes, stretching out from their sides—which chariots were driven with great force through the ranks of the enemy, mowing them down in their course—it does not invalidate the statement that iron was abundant, and used in great quantities by the nations of that early period. "And the staff of his spear was like a weaver's beam, and his spear's head weighed six hundred shekels of iron " (1 Sam. xvii. 7). "But the man that shall touch them must be fenced with iron, and the staff of a spear " (2 Sam. xxiii. 7), probably referring to coats of mail. "And Zedekiah, son of Chenaanah, made him horns of iron " (1 Kings xxii. 11). "He shall flee from the iron weapon, and the bow of steel shall strike him through" (Job xx. 24). Here we have not only iron weapons, but steel; and if we allow the use of steel bows for shooting of arrows, it indicates an extraordinary advance in the manufacture of iron as well as implements of warfare. That such instruments as iron-tipped arrows may have also been used is indicated by the following passage— "Canst thou fill his skin with barbed irons? or his head with fish spears?" (Job xli. 7).

There are again many references to its use in arts and manufactures. Job speaks of having

his words graved in the rock with an iron pen.
" And he brought forth people that were therein,
and put them under saws, and under harrows
of iron, and under axes of iron " (2 Sam. xii. 31).
" And David prepared iron in abundance for the
nails for the doors of the gates, and for the join-
ings " (1 Chron. xxii. 3). " The sin of Judah is
written with a pen of iron " (Jer. xvii. 1).
" Thou hast broken the yokes of wood, but thou
make for them yokes of iron " (Jer. xxviii. 13).
" They drank wine, and praised the gods of gold
and of silver, of brass, of iron, of wood, and of
stone " (Dan. v. 4). " For three transgressions
of Damascus, and for four I will not turn away
the punishment thereof, because they have thrashed
Gilead with thrashing instruments of iron "
(Amos i. 3). " As iron sharpeneth iron, so a man
sharpeneth the countenance of his friend " (Prov.
xxvii. 17). " And he shall cut down the thickets
of the forest with iron " (Isa. x. 34).

The mode of working and tempering iron seems
to be referred to in the following passage—" The
smith with the tongs both worketh in the coals, and
fashioneth it with hammers, and worketh it with
the strength of his arms " (Isa. xliv. 12). Michaelis
translates the passage thus—" The smith bends the
iron, works it in a fire of coals, and forms it with
the hammer; he labours on it with a strong arm."

We have iron referred to also in common use for
domestic purposes. " For only Og, king of Bashan,

remained of the remnant of giants; behold, his bedstead was a bedstead of iron " (Deut. iii. 11).

Indeed iron is referred to in various ways in
Scripture, proving without doubt not only its
general use among the nations, but their understanding of its properties and its manufacture; so
that the poetic and moral writers make its properties figures for illustrating their lessons. There
are iron bars and pillars and gates and walls
referred to as impenetrable. Different qualities of
iron are referred to by Jeremiah—" Shall iron
break the northern iron and the steel?" " Foras-
much as iron breaketh in pieces and subdueth all
things, and as iron that breaketh all these, shall it
break in pieces and bruise " (Dan. ii. 40).

The possession of iron seems to have been
considered as a source of wealth to a nation, and
is referred to as such along with other metals by
Joshua—" And when Joshua sent them away
also unto their tents, then he blessed them and
spake unto them saying, Return with much riches
to your tents, and with very much silver, and with
gold, and with brass, and with iron" (Joshua xxii. 8).

Particular attention has been given to the
references in Scripture, because the existence of
iron in large quantities in ancient times has been
doubted, and the opinion generally maintained
that bronze was used previous to iron, and for all
the purposes for which iron was afterwards used.
It is also stated by Greek writers that iron was

discovered about the year 1406 B.C.—fifty years after Moses wrote. We are rather inclined to differ from these opinions; and the more they are considered in connection with all the circumstances, the more the assertion that bronze was used instead of iron for all purposes appears incorrect and untenable. The first notice of bronze in the Scriptures is in connection with iron; and we have shown that so far as Scripture history is concerned, iron is generally named along with bronze. Even allowing that Moses, in stating that Tubal-cain was an instructor of every artificer in iron and brass, was only repeating a common tradition of his day, it will not affect the correctness of the assertion that iron is named as early as bronze. The book of Job has many references to iron, showing that it was used for those purposes to which Greek writers say bronze was applied long after. It is difficult to understand how bronze-edged tools alone could be used by the Israelites for engraving the twelve precious stones set in the breastplate, these stones being of the hardest known substances. This is admitted to be difficult to explain by those who have considered the subject. Dr Lester, in a recent paper in the " Philosophical Transactions," thinks that the ancients possessed the secret of tempering steel better than the moderns. This would explain the difficulty.

In Rawlinson's " Five Great Monarchies," speaking of the metals in use in the reign of Urukh, a thousand years before the engraving of the stones for the breastplate, there is mention made of iron being in common use along with bronze, copper, and lead, and also stone implements. In what metallic age would some of our archæologists place King Urukh?

We have also abundant references in profane history to the prevalence of iron in ancient times to verify the statements in Scripture. Pliny, and several other ancient writers, mention various countries and places which in their time produced excellent steel. The art of hardening iron and steel by immersion while red-hot in cold water, is referred to by Homer. He says, that when Ulysses bored out the eye of Polyphemus with a burning stake, it hissed in the same manner as water when the smith immerses a piece of red-hot iron in order to harden it. The passage is thus translated by Pope—

> "And as when armourers temper in the ford
> The keen-edged pole-axe, or the shining sword,
> The red-hot metal hisses in the lake—
> Thus in his eyeball hissed the plunging stake."

In ages near to the Christian era, among the Greeks and other nations the operations in the manufacture of iron were performed by slaves. If this was the practice in ancient Egypt, it is not improbable that one of the forms of slavery used to coerce the Israelites was labouring at the iron

furnaces; and when Moses speaks of having
delivered them from the iron furnace, there may
have been a reality in the reference which would
be fully understood by all who had suffered such
oppression.   Diodorus, writing long after the
exodus, represents the labour of the slaves at the
iron furnaces of Greece as the most intolerable of
all tyrannies.

In order to give some idea of the extent of the
manufacture and use of iron among what is termed
the Gentile nations in the earliest ages of their
history, we may refer to Hesiod, who is amongst the
earliest authors whose writings have come down to
us.   He divides the ages of man into four, namely,
golden age, silver age, bronze age, and iron age—
which last was the one he lived in—supposed to
be about 944 B.C., a few years after Solomon's
death.   Although this is no doubt an allegorical
comparison, and we think has not the same meaning as our archæological division into stone,
copper, bronze, and iron ages as descriptive of
the ascending progress of civilisation, at the same
time it must have a certain relation to historical
fact, and is therefore significant in this inquiry
as iron must have been in pretty general use in
Hesiod's time, so as to assume in the poet's mind
such a position.   Homer represents one of his
heroes encouraging his men in the following
words—"Their flesh is neither stone nor iron,
to endure strokes given with the cutting edge of

bronze;" rather an anomalous circumstance to use a bronze weapon, when iron is acknowledged to be able to resist its strokes, and to be every way superior; and as Homer flourished in the iron age of Hesiod, one would have expected his hero to have been furnished with iron weapons; while Achilles' sword is stated to have been made of bronze. The Right Hon. W. E. Gladstone, in his "Juventus Mundi," places Homer in the copper age of the archæologists, while Homer himself makes many references to the ordinary use of both bronze and iron. Dr Moore, in his "Ancient Mineralogy," says, "In Homer's time iron was well known, and the poet employs, as well as Moses does, the names of iron and brass in both a literal and figurative sense, as though the two metals were applied indifferently to the same purposes, whether of war or peace, and regarded as possessing in like manner the properties of hardness and tenacity. Iron, however, was used much more sparingly than brass, being for the reasons no doubt that have been stated, much the rarer of the two; accordingly we find that a rudely cast mass of iron, which had been used by Eëtion as a quoit, is represented by Achilles when he offers it among the prizes at Patroclus' funeral, as a five years' provision of iron for one who cultivates extensive and rich fields, so that his shepherd or his ploughman when they need iron, may, without going to the city for it, satisfy their wants."

o

All this is very suggestive. Iron is mentioned
as an article of commerce when Mentor tells Telemachus that he has iron which he is going to
exchange at Tennes for copper. It may be that
these were iron tools or iron to make tools for the
miners who were working in the copper mines.
Homer refers to iron being in use for various purposes. " He applied the string to his breast, and
the iron to his bow." Areithous "broke the phal-
anxes with an iron club;" probably this was something like the spear of Goliath with its head of
iron. The chariot of Juno is said to have had an
iron axle with brass wheels. It is also mentioned
as spoil taken in war, and as treasure laid up by
the wealthy. In a similar way it is referred to in
Scripture—" But hither I shall bring gold and
red brass, and women elegantly clad, and white
iron." " Many precious treasures be among my
fathers' riches, brass, and gold, and iron curiously
wrought." It is also referred to by Homer in a
figurative sense: " The iron tumult reach the
brazen heavens," and " the insolence and violence
of suitors to reach the iron heaven." Ulysses
is said to be made of iron, because he does
not know fatigue, and so on. Iron is referred
to by Homer in the same style as it was centuries
before his day by Moses and Joshua; so that it
is rather anomalous to represent Homer as living
at the commencement of the copper age of our
archæologists while he is so familiar with iron.

Speaking of Egypt, Mr Basil Cooper, in his paper upon the "Antiquity of the Metals," says, "We find that there also, as well as on the classical soil of Greece and Rome, the origin of the art of working in iron is pushed back into the mythological and prehistoric age. We have no reason to doubt the testimony of Diodorus, when he repeats that the Egyptians assigned this invention also, as well as all the other more important arts of life, to their great national culture-divinity, Osiris."

In "Iron" it is stated that an English gentleman has recently discovered near the Wells of Moses, by the Red Sea, the remains of ironworks so vast that they must have employed thousands of workmen. Near the works are to be found the ruins of a temple, and of a barracks for the soldiers protecting or keeping the workmen in order. These works are supposed to be at least three thousand years old.

In Layard's "Ancient Nineveh," he says, "When descending the Tigris on a raft to the ruins of Nimroud, he passed a considerable cataract, formed by a solid wall of masonry, which wall his guide told him was the remains of a great dam built by Nimroud, and that in autumn, before the winter rains, the large stones of which the dam was constructed were united by clamps of iron, which were frequently visible above the surface of the stream. It was, in fact," says Layard, "one of the monuments of a great people to be found in

all the rivers of Mesopotamia, which were undertaken to secure a supply of water to the innumerable
canals spreading like net-work over the surrounding country, and which, even in the days of
Alexander, was looked upon as the works of an
ancient nation." We would like to know by
what means the iron was fixed into the stone.

As a proof of the plentiful use of iron in very
ancient times among the Assyrians, we give the
following extract from Vaux's " Nineveh ": " In
one of the chambers there was a large quantity of
iron amongst the rubbish, and Mr Layard soon
recognised in it scales of armour, similar to that
on some of the figures. Each scale was, when separated, from two or three inches long, rounded at
one end and square at the other, with a raised or
embossed line in the centre. The iron was so
much decomposed that it was difficult to detach
it from the soil. Under the earth other portions
of armour were found, some of copper, others of
iron inlaid with copper. A perfect helmet was
found, resembling in shape and in the ornaments
the pointed helmet of the bas-relief, but it fell
into pieces immediately the earth was removed.
The lines which are seen round the lower part of
the pointed helmet in the sculptures are thin
stripes of copper inlaid in iron." These extracts
are very suggestive, in a practical point of view,
not only as indicating a common use of iron, but
also great skill in the working or manufacturing

of articles in it. The inlaying of one metal with another, and that the hardest of the metals, shows a degree of skill not inferior to anything we do in our day in the working of metals, and it also goes far to illustrate many references made in Scripture to metals and metallic implements, which would otherwise be obscure.

It seems to us that in all ages of the world, so far as history and discovery has gone, there has always been some active nucleus or centre of civilisation, and in these centres the arts were known to the greatest perfection of the age, and in most cases where researches have been made, when one metal has been found, there is not wanting evidence of the presence also of others. It may be that a metal, such as copper, where it is found native, may be in use where there is no bronze; but it is almost impossible to believe that any people should obtain such perfection in such a complicated process as the manufacture of bronze, and not have iron, the ores of which being abundant and wide-spread, and the metal as easily reduced from the ore as either tin or copper, even without the use of twenty pairs of bellows. In our own forefathers' time when they manufactured iron in bloomeries, the fire was blown by the winds of heaven. Hence, we think that iron was known as early as bronze, although from its requiring much more labour to fit it for useful purposes, and from other causes, it may not have been so

abundant in some localities as bronze was; so that
it may have been used along with bronze for purposes to which the latter could not be applied
alone, such as hewing or engraving on hard stone;
in which case, as Wilkinson suggests, a casing of
iron may have been used, and, indeed, tools cased
or tipped with iron have been found in our own
country. The fact that Sir G. Wilkinson found a
bronze chisel in a quarry in Thebes, is not sufficient evidence that they either quarried or hewed
stones with such, more especially when it is known
that at the time when this chisel was used, iron
and steel were in common use. Butchers are
represented on the Egyptian monuments as sharpening their knives with steel, as butchers are in
the habit of doing at the present day, and as
seems to have been the practice in Solomon's
time, when he makes it a symbol for stimulating
intercourse between man and man to sharpen
their understanding and produce warmer friend-
ship. We hardly think that the knives used by
the Egyptian butchers were of bronze. About
the same time that the chisel is said to have
been used, there occurs a significant passage in
the writings of Moses—"And there shalt thou
build an altar unto the Lord thy God, an altar
of stones: thou shalt not lift up any iron
tool upon it. Thou shalt build the altar of the
Lord thy God of whole stones" (Deut. xxvii. 5,
6). And again, "As Moses the servant of the

Lord commanded the children of Israel, as it is written in the book of the law of Moses, an altar of whole stones, over which no man hath lift up any iron" (Joshua viii. 31). This prohibition is against hewing or dressing stones for the altar. If bronze was used for this purpose, the prohibition is not only incomplete but meaningless. As the Israelites had just left Egypt, we think the use of iron for hewing stones, referred to in this passage, is applicable to the Egyptian practice, and if, as A. G. Wilkinson suggests, a shield of steel may have been used over the bronze edge, it is an evidence that iron was in use as early as hewing of stones, and as early as bronze ; and to tip bronze with steel for hewing must have been a very nice operation, and shows the skill possessed by the ancients in working in the metal. Bronze tools so tipped have recently been found. Such facts go far to show that the regular succession of stone, bronze, and iron, so dogmatically laid down by some archæologists, is not a universal law; indeed, there is evidence to the contrary in our own day in Cape Colony, where there are savages who appear never to have known or wrought in copper or bronze, who, nevertheless, are most excellent manufacturers and workers in iron.

Overlaying of iron with tin has been mentioned as an art probably known to the ancients. Pliny mentions the overlaying of copper with tin, as is now done, to protect culinary vessels from poisonous verdigris. No articles made of iron, overlayed with tin, have been found; but this can scarcely be a proof against the use of tinned iron in ancient times, as it is very easily destroyed by corrosion. It is now known that iron was covered with brónze, which gave the strength of iron when required, and, at the same time, retained the external beauty of the bronze. Several objects having this overlaying were found in the ruins of ancient Nineveh by Mr Layard, who has the following passage upon this remarkable fact: " It would appear that the Assyrians were unable to give elegant forms or a pleasing appearance to objects made in iron alone; and that, consequently, they frequently overlayed that metal with bronze, either entirely or partially, by way of ornament."

It may be remarked here, that the operation of covering iron with bronze is one of considerable difficulty in our day, and not known to have existed until these examples of the ancients were brought to light. These facts prove that the ancients were well skilled in metallurgical operations, and verify the truth of the words of Scripture in their many references to iron and other metals. Perhaps, however, some of these references allude to practices now unknown; hence the difficulty often experienced by commentators and others who endeavour to explain them.

LEAD.

LEAD is seldom found in the earth in a metallic state, and then only in very small quantities, but it is found widely diffused, and in great abundance, as an ore combined generally with sulphur. The ores of lead are all heavy; and are white, green, or slate coloured. Their great weight could not fail to attract attention, and consequently we find lead noticed at a very early period.

The method of extracting the metal from the ore is somewhat similar to that described for copper and tin. The ore is generally found in veins of rocks, and has to be dug out in masses, which are often mixed with earthy matters. These masses are broken into small pieces, and then crushed fine, and washed in a stream of water, which carries away the greater part of the earthy matters, owing to their being lighter than the ore. When the ore is an oxide or carbonate, it is mixed with coal, or other carbonaceous matter, and heated to redness in a furnace; the metal is then obtained at the bottom of the furnace or vessel in which it had been heated.

The principal ore of lead at the present day, and probably also in ancient times, to judge from the refuse mounds left behind, is the sulphide (galena), which is of a blue slate colour. This is subjected for a time to a low red heat, in a free current of air, by which a portion of the sulphur is oxidised, part being carried away as sulphurous gas, and part remaining as sulphate of lead. The heat of the furnace being then increased, this sulphate undergoes decomposition, and reacting on the remaining sulphide ore, all the lead is reduced to the state of metal without any mixture with coal. This is done in a reverberatory furnace, and is the process by which lead ore is smelted at the present time.

The sulphide of lead may also be reduced by putting it into a furnace or crucible, with flux—such as soda or potash—or mixing with it metallic iron. These last processes, however, are not profitable, and not applied on a large scale. As to whether any of these methods were adopted by the ancients for obtaining the metal from the ore we have no reference, either in sacred or profane history, except the obscure notice of Pliny already referred to.

Dr Kitto, in his "Physical History of Palestine," says, "that lead ore is known to have existed in considerable quantity in the neighbourhood of Sinai, and also near Egypt, hence the supply to Egyptians and Israelites was of easy access, and is quite consistent with the early notice of it in Scripture."

In Moses' triumphal song, celebrating the overthrow of Pharaoh in the Red Sea, he says, "They sank as lead in the mighty waters" (Exod. xv. 10), a familiar expression, the same as our saying, "As heavy as lead," &c. ; an expression that clearly proves that lead was then in common use, and its common physical property well known. It is also mentioned among the spoils taken by the Israelites from the Midianites—"The gold and the silver, the brass, the iron, the tin, and the lead" (Numb. xxxi. 22), evidence, we think, that it was held of considerable value among the Hebrews. In the book of Job, the patriarch makes a beautiful reference to a use to which this metal was applied, "Oh that my words were now written! Oh that they were printed in a book! that they were graven with an iron pen and lead in the rock for ever!" This passage clearly indicates that when the book of Job was written, lead was not only well known, but applied to one at least of the fine arts. This reference is considered to refer to an inscription first engraved on stone, and afterwards filled in with lead—a beautiful application for the preservation of writing. The various objects for which the metal lead was used and applied in very ancient times are not mentioned in Scripture. Its extreme softness, the rapidity with which it tarnishes, its want of sound, and other general properties, are against its employment for domestic purposes, but we believe it was used for many of the purposes to which it is put at the present time. Pliny mentions that it

was used in his day, and before his time, for water
pipes, as it is at present, and that sheet lead was
used for covering articles for their protection.   He
also mentions it being used for writing tables.
Whether, however, it was applied to these objects
by the Egyptians or Hebrews in the earlier ages
of their history, is not certain, but we think it probable that it was used as early as the days of Moses
for such purposes.   In order to show the extent of
its use, and the enormous quantity possessed in
ancient times, but later than that of Moses, we
quote the following from Kitto's " Cyclopædia
of Biblical Literature," article, "Babylon," in
reference to the stupendous hanging gardens of
Babylon in the time of Nebuchadnezzar:—

" The level of each terrace was then formed in
the following manner: The top of the piers was
first laid over with flat stones, sixteen feet in
length, and four feet in width; on these stones
were spread beds of matting, then a thick layer of
bitumen, after which came two courses of bricks,
which were covered with sheets of solid lead."

We extract the following from Dr Thomson's
" Land and the Book " :—

" In the wildest of those gorges whose outlines
lie in misty shadows along the south end of
Lebanon, bursts out a copious spring called Neb'a
et Tāsy,—Fountain of the Cup.   It is the source
of the Zahrany.   The ancient Sidonians coveted
this ice-cold water, and did actually lead it to their

city, along a line of canal which might well con-
found the boldest engineer. A channel was hewn
in the rock, into which the new-born river was
turned, and thence carried down the gorge south-
ward until it could double the promontory of
Jerju'a, after which it meandered as it could north-
ward for eight miles, spanning deep ravines over
high arches, and descending into Wady Kefrah below Jeba'ah. Beyond this, the aqueduct was led
along frightful cliffs, where goats can scarcely keep
their feet, for more than a mile, and thence it
followed the ridge of Kefr Milky, past the village,
into the wady of the Sanîk, where it was joined by
another aqueduct from Neb'a or Râhib, the source
of that river. The two canals were taken thence
down the river, but separately, one about fifteen
feet above the other. The system of arches by
which these works were carried across the ravines
and rivers is still almost perfect, and the cliffs to
which they cling are absolutely perpendicular for
miles together. As there are no traces of arches
by which the water was led across the low plain
up to the city, it has been conjectured that the
Sidonian engineers were acquainted, at that early
age, with the principle in hydrostatics that water
will rise to the level of its source. People also
tell me that fragments of earthen pipes, incased in
lead, have been dug up in the gardens in the probable line of these canals. These may have served
to conduct the water to the city."

There is much that is suggestive in the above extract; the incasing of earthen pipes in lead for the conveyance of water agrees with what Pliny says in reference to covering articles with sheet lead for their protection. But lead pipes were used as conduit pipes in ancient Rome. Among the metallic articles found in Pompeii are lead pipes that had been used for conducting water through the city. James Young, Esq., F.R.S., after visiting the ruins of Pompeii, sketched for us the form of these pipes in section Ⓑ, the opening at *a* being filled in with lead to complete the pipe. In order to find whether the metal used for filling in partook of the nature of solder, a small piece was analysed and found to be pure lead, no tin being present. It would be important to know if other ancient pipes were made in the same way. This form of pipe shows that in these days they did not practise our method of drawing out lead pipes; but whether the lead was cast or beaten into sheets, and then beat into shape upon a mandril, we do not know. It is probable that when the sheets were beaten up into the form of a pipe as shown, melted lead was poured into the space at *a*, completing the pipe. In this case, a mandril or other means would be required inside, to prevent the melted lead from passing through.

Lead ore is found abundantly in Britain, and there is every evidence that the lead mines of this country were wrought by the ancient Romans, and probably lead formed a part of the traffic of the Phœnicians. Sir R. Murchison says, that "lead cast in Roman moulds, *pigs*, in fact, of the age of Hadrian and other emperors, have been found in Fifeshire, Derbyshire, Yorkshire, and some other counties. The shape of the ingots or pigs are nearly the same as at the present day, and the inscriptions are made in raised letters on the top."

In writing of these, Mr J. Phillips has the following passage :—

" A third with the inscription, also in raised letters on the top, was found on Matlock Moor, in the year 1787. It weighed 173 lbs., and was seventeen and a half inches in length, and at bottom twenty and a half. The inscription was, TI. CL. TR. LVT. BR. EX. ARG."

Under " silver " we have shown that lead was used in the earliest times for purifying silver in the same manner as we do it now—namely, mixing lead with silver and melting them upon a cupel, or flat vessel made of bone earth, and then oxidising the lead by blowing upon the surface of the fused mass, till the silver is left pure. Having established the fact that lead was used for purifying silver, and knowing that the greater portion of lead ores contain silver, sometimes in considerable quantity, and that lead ores are in the present day an important source of that metal, we are warranted in believing that this fact was known to the ancients, and that their lead ores were a source of silver to them, and may have been an important source. Pliny states that the ore of lead is found mixed with silver in the same mine.

It is highly probable that the Phœnicians were extensively engaged in lead smelting, but whether it was smelted at the mines, and imported as metal into Phœnicia as tin was, we do not know, but the prophet Ezekiel mentions it as an article sold in their markets thus: "With silver, iron, tin, and lead, they traded in thy fairs" (Ezek. xxvii. 12).

The familiar tone in which lead is always referred to in Scripture, and by ancient authors, lead us to believe that it was in use in the earliest ages.

In the purifying of silver with lead, they necessarily would form litharge, or oxide of lead. This oxide, when heated with coal or carbonaceous matters, produces metal again—a process which would undoubtedly be practised for recovering the lead. This oxide of lead is used in the present day for the manufacture of salts of lead, and for paints, glazes, enamels, and making glass. Some of the glazes upon articles of pottery, and the paints on the clay vessels found in ancient Egypt and Nineveh, have been analysed, and found to contain oxide of lead ; in other words, to be composed of the same materials as those which we now use for the same purposes, and which, until lately, was considered an invention of our own time. Different coloured glazes and pigments used by the ancient Assyrians have been found by analysis to contain oxides of copper, cobalt, iron, tin, and lead.

Lead is also used by us along with tin, for soldering certain metals together. The Scripture does not mention the composition of solder, yet the fact that soldering was practised, is evident from the following passage : " So the carpenter encouraged the goldsmith, and he that smootheth with the hammer him that smote the anvil, saying, It is ready for the soldering : and he fastened it with nails, that it should not be moved " (Isa. xli. 7). This is another beautiful description of a practical process in metallurgy, and is similar to the method now practised in soldering two pieces of metal; after being smoothed and prepared, they are fastened together by iron clasps, or holders for soldering. The word, nails, according to the margin, is a mistranslation—meaning holder or clasp. It is also a fine description of the division of labour—a system not only economical, but calculated to make the finest work, and the most expert workmen. The nature or kind of solder is not stated in the passage, but the term goldsmith being used, indicates that the article being soldered may have been of gold or silver. Solder used for these metals is generally made of copper, silver, and tin, or such an alloy as melts at a lower heat than the metals soldered. The passage, however, states the practice, and

P

indicates a general use, and this has been verified by articles found in ancient Egypt that had been soldered. Wilkinson says : " In coarser work, or in those parts which were out of sight, the Egyptians soldered with lead. The oldest specimen of metal soldered which I am acquainted with, is the sistrum of Mr Burton—its date is uncertain ; and though, from the style of the figures engraved upon it, we may venture to ascribe it to a Pharaohnic age, the exact period when it was made cannot be fixed." Pliny states that lead was used for soldering lead and other metals, and this agrees with the manner in which the lead pipes found in Pompeii were made. It must be borne in mind that the term solder, both in ancient and modern times, does not indicate any particular compound of metals, but simply a metallic substance, used for joining metallic surfaces. To a plumber the term means an alloy of lead and tin, to the coppersmith an alloy of copper and zinc, to the silversmith an alloy of copper, silver, and tin; so that when solder is mentioned in ancient writings, as in modern, it does not necessarily mean any particular alloy.

MERCURY.

WHEN the metal mercury, or quicksilver, was first discovered, is not known. But that it was known in ancient times, at least as early as Aristotle's time, is clearly shown by Pliny and other writers. There are some, however, who conclude from the mysterious language used by the adepts or alchemists in ancient times, and the ancient ideas regarding metals, more especially mercury—which was looked upon as the mother by which all the metals were fructified, purified, and brought forth—that this metal may have been known to the ancient Egyptians in the earliest ages, and also to the Hebrews. Such writers consider that in the passage where Moses directs that all the metals taken from the Amalekites should be made to pass through the fire, and afterwards to be " purified by the water of separation," that this water of separation was mercury. We quote the following on this passage from Sir John Petus' translation of the works of Lazarus Erckern, 1683 :—

" And we are assured that in Moses' time, they had the knowledge of all metals, as may be read in Numbers xxxi. 21, where Moses taught the soldiers how the spoils of their *heathen enemies* were to be purified, commanding (as from God) that all their gold, silver, brass, copper, iron, tin, and lead, and everything that endureth the fire (in the furnace, according to the Syriac), should be purified by fire, and then to be accounted clean. Yet it is also said in that text that it shall be separated by the water of separation—by which water certainly is meant quicksilver—because this doth *purify*, *cleanse*, and *devour metals*, and, as Dr Salmon calls it, a *volatile juice* or *liquor ;* for nothing but fire, or that *quicksilver, aquafortis*, can separate those metals. The water of purification of men was a distinct water from the water of *purification* and *separation* for metals, and the ingredients of the one are communicated to us ; but the Holy Spirit thought fit to conceal the other from us."

We think there is no foundation for this supposition of this old master of assaying and metallurgy, Lazarus Erckern. The water of purification was prepared by burning a red heifer entire, then collecting the ashes and mixing them with water, as given in Numbers xix. ; besides mercury does not purify, nor can it be used for separating the metals named in the passage. There is no reference to the metal mercury, so far as we can discover, in the Old Testament Scripture, and we think that from its peculiar qualities of dissolving, or apparently absorbing and devouring other metals, had such a metal been known to the poets of ancient Israel, they would have used it as a figure descriptive of some of the conquering nations. The property that mercury has for dissolving and amalgamating with metals, particularly all those known to the ancients, except iron, would very soon suggest an important use. It does not unite with earths; when therefore gold or silver is diffused through earthy matters, such as sand, and quicksilver be added, it combines with the metals, and forms an amalgam or solution, taking up the precious metals from the sand, which is easily removed from the amalgam. As mercury is volatilised at a moderate temperature, this amalgam has merely to be heated, when the mercury distils off, and leaves the gold or silver behind; if both metals be present they are mixed. That this use was made of quicksilver by the ancients is stated by Pliny, who says that it was used for separating gold from its impurities, and that the quicksilver was afterwards separated by straining through leather, a process still in use for separating the excess of mercury that may have been used, but it requires heat to separate the last portions.

" Vitruvius," says Beckmann, in his " History of Inventions," " describes the manner of recovering gold from cloth in which it has been woven, 'the cloth,' he says, 'is to be put into an earthen vessel and placed over the fire, in order that it may be burned. The ashes are then thrown into

water, and quicksilver added; the latter attracts
the particles of gold and unites with them; the
water is poured off and the residue put into a
piece of cloth, which being squeezed with the
hands, the quicksilver, on account of its fluidity,
oozes through the pores, and the gold is left pure
in a compressed mass.'" The gold would not be
obtained *pure* by this process, but would contain
some of the quicksilver. The heating of the mass
in a crucible, or any suitable vessel over a fire,
however, would drive off as vapour the remaining
mercury, and pure gold would then be obtained.
Probably they did not know in Vitruvius's time
the art of distilling in close vessels, and recovering the mercury. When gold and quicksilver
amalgam is rubbed upon other metals till they
are coated with it, such as, for instance, an article
made of silver, the coated silver then heated, the
mercury is volatilised, and the gold remains upon
the surface of the silver. By this means silver
and other metals were gilt. This process of gilding was extensively practised in ancient times,
but how far back we do not know, and it has
continued to be practised up to the present time;
indeed it was the principal process for gilding till
1840, when it was to a great extent superseded by
the electro process. Other uses to which quicksilver was applied by the ancients are not very
clearly defined. At whatever period mercury was
known as a metal, its principal ore, the bisul-

phide, wherever found, would not fail to attract
the attention even of the rudest people by its
brilliant red colour, and there are sufficient
evidences to show that it was used as a pigment
or paint in very early times, under the name of
vermilion. Pliny says it was used as a pigment
by the Romans, and was in ancient times held in
high estimation, and used for religious purposes.
In ancient Rome it was the custom on festival
days to colour with vermilion the face of the
statue of Jupiter, and the bodies of triumphant
generals. Amongst the Ethiopian nations also
it was held in great esteem, their nobles being
in the habit of painting their bodies all over with
it; their statues also were coloured vermilion. It
was the colour of honour, which will be more
fully treated under dyeing. Vermilion was also
used as a paint for rooms amongst the Jews. Jeremiah, in chapter xxii., referring to the luxurious
dwellings, says (verse 14), " I will build me a wide
house and large chambers, and cutteth him out
windows; and it is ceiled with cedar, and painted
with vermilion." Ezekiel also refers to its use for
paint in Babylon thus (xxiii. 14.), " For, when she
saw men pourtrayed upon the wall, the images
of the Chaldeans pourtrayed with vermilion." We
can hardly conceive of an extensive use of this
ore, without a knowledge of the metal, as it is
obtained from the ore at a heat under redness; so
that we think it possible that at whatever date we

can trace the use of the ore, the metal would
then be known.   However, in this inquiry we
must guard against being led astray by the name
given to this paint by ancient writers, as this
name sometimes only refers to the colour, whatever it is made of, and in this way is confusion.
This ore of mercury is sometimes termed minium,
and ranked with this ore of lead, and even with
the red oxide of iron, the colour being similar,
so that it is often difficult to know what the
writer refers to, unless when some peculiar property of the paint is given; but this is not attri-
butable, we think, to ignorance concerning its
distinctive character on the part of practical
men, but to ignorance on the part of Pliny and
other writers, who had no technical knowledge.
Many of our own historians make sad mistakes in
technical matters, which is not to be wondered at;
but what is really surprising is, that some of the
leading encyclopædias and dictionaries are not free
from blemishes in this direction; for instance, the
metal zinc is spoken of in some dictionaries as a
sort of *semi-metal*, and ranked with sulphur.

ZINC.

THE metal zinc was discovered in the sixteenth century of our era. It exists abundantly in nature in a mineralised condition, forming ores, some of which were well known to the ancients. Its alloy with copper forms what is now technically named brass, and which is said to have been discovered in the thirteenth century. This discovery came no doubt from finding that a certain mixture of what was a zinc ore and carbon, with copper in-fusion, gave a uniform compound metal, capable of being used in the arts. At the same time, there is no doubt but the alloy of copper and zinc was known to the ancients, at least in Pliny's time. Its oxide was known and used in his day as medicine, and was obtained from the copper furnaces in the island of Cyprus. Calamine, or carbonate of zinc, is often mentioned as having been found in Cyprus.

Brass is easily formed by fusing together such ores of zinc with carbonates of copper, and might be so obtained by the ancients long before the discovery of zinc as a separate metal. Zinc being a volatile metal would make the discovery of its true character longer of being found out, and when it is volatilised in an open furnace it passes off as an

oxide. It was this sublimed oxide, condensed in
the colder portions of the copper furnaces, that
was used in medicine. That an alloy of zinc
and copper was known and used for castings, is
proved by the analysis of several. The following
two are by Mr J. A. Phillips :—

Brass of Cassio family B.C. 20.

Copper,	.	.	.	82·26
Zinc,	.	.	.	17·31
Iron,	.	.	.	·35
				99·92

Large brass of Nevo A.D. 60.

Copper,	.	.	.	81·07
Zinc,	.	.	.	17·18
Tin,	.	.	.	1·05
				99·30

The name calamine, or cadmia, applied to this ore
of zinc, is said to have been derived from Cadmus,
who lived about 1500 B.C., and who is said to have
introduced the making of brass at Thebes, by mixing the ore with copper in a fire. We have doubts
of the correctness of these statements. If brass
had been made in Thebes in that early age, we
could hardly fail to get some of the remains of
brass articles among the ruins of that ancient
city, but we have not seen or heard of anything
approaching to brass, nor have we seen the analysis
of any alloy having zinc in its composition until we
come near to the Christian era. It will be exceed-

ingly interesting if further explorations should
bring forth articles made of brass or having zinc
in them, as such might indicate the source of the
copper or its ores.   It is quite possible that calamine may have been imported into Thebes for a
time, and used as stated above, and afterwards,
from certain causes, ceased to have been brought
into the country; so that the brass made from it
may not have been very extensive, and the absence
of brass articles among the findings may not be
an absolute objection to the statement referred to.
However, we must wait further inquiry.   We have
no scriptural evidence of any such thing as brass,
as we understand the term, having been known
among the Jews.

ANTIMONY.

WE do not think that the metal antimony was known to the ancients; but one of its ores, the sulphide, was known from a very early date. This was termed stibium, and was used by women as a paint for painting the eyebrows, eyelashes, and edges of the eyelids, in order that the eyes might appear larger and brighter; a large bright eye being considered a point of beauty in a woman. The practice is still continued by Eastern women, and we find it repeatedly referred to in Scripture. In 2 Kings ix. 30, it is said, " And when Jehu was come to Jezreel, Jezebel heard of it; and she painted her face " (translated eyebrows, or she put her eyes in painting, as in the marginal notes), " and tired her head, and looked out at a window." Jeremiah says (iv. 30), " Though thou clothest thyself with crimson, though thou deckest thee with ornaments of gold, though thou rentest thy face " (eyes in the marginal rendering) " with painting, in vain shalt thou make thyself fair." And Ezekiel refers to the practice amongst Jewish women thus (xxiii. 40): " And furthermore, that ye have sent for men to come from far, unto whom a messenger was sent; and, lo, they came: for

whom thou didst wash thyself, paintedst thy
eyes, and deckedst thyself with ornaments." Commentators agree that the word translated "painting," or "painted," should be stibium; however,
some of them consider that the painting was done
with plumbago or black lead, and speak of this as
a sort of lead ore. Stibium was the name for the
sulphide of antimony. Plumbago is a carbide of
iron—that is, a combination of carbon with iron—
and has no relation whatever to lead or its ores.
Sulphide of lead, or galena, may also have been
used for the same purpose, for when finely ground
it will stain the skin black, but not so deeply as
either of the others. Probably all three were used
according to circumstances, the wealthy using the
best, which was the stibium.

Pliny refers to stibium, but in a way which
makes it doubtful if at all times he means the pure
sulphide of antimony. Dioscorides, in describing
the preparation of the paint, says, " That it is to
be put into a lump of dough, and that put into
a fire, and covered over or buried in the coals, until
it is reduced to a cinder; it is then extinguished
with milk and wine, and again placed upon coals,
and blown until ignition; if burned longer it
becomes lead." Pliny's description for the preparation of the paint is nearly the same, only he
suggests the use of cow dung instead of dough,
and cautions moderation in burning, lest it should
be converted into lead. From these remarks it is

probable that the ancients may have had metallic antimony reduced during these operations, but mistook it for lead, which it resembles in appearance, and thus its distinctive character as a metal may have been unknown. We think it evident from the various applications to which the ores of antimony, zinc, arsenic, &c., were put by the ancients, that these metals must frequently have been accidentally reduced to reguline condition, but that the ancients, having a settled physical dogma that there were only seven metals, these extra metals, when so reduced and noticed, were considered, first, on account of the impossibility of there being more metals than seven, and second, on account of their resemblance to lead or tin, to be one of these metals in an impure and valueless condition. These conventional ideas prevented, till modern times, the discovery that antimony is a distinct metal. Some of its preparations, as the oxides, were used by the discoverer as medicines, and these were given to the monks in the monastery in which he was a member, to test their effects, which often produced serious and fatal results, hence the name given to the metal, *anti-monk* or *antimony*.

DYEING.

RECORDS and relics of decorative art supply interesting data, from which may be deduced important conclusions concerning the degrees of refinement in civilisation attained by ancient nations. Among such arts dyeing holds a forward place. Being curious to know what were the processes followed by dyers in ancient times, we made patient search among such ancient literature as was accessible to us, and which seemed at all likely to contain the information wanted, but regret to say, found no reliable, nor even intelligible, description of such processes anywhere. We are, therefore, compelled to return to the deductive and analytical course which we pursued when treating of the metallurgical arts, and deduce our conclusions from such recorded analyses as we have found of ancient dyed fabrics, and also from the colours and dyeing agents which we know the ancients were acquainted with.

We have divided the subject into three heads. *First*, The fabrics or materials dyed; *second*, The colouring matters used; and *third*, The colours dyed. A knowledge of the fabrics, and the

colours they were dyed, is of the utmost importance in this inquiry; for the same operations,
often the same dye drugs, do not produce the same
effect upon different fabrics.

Skins of animals.—Dyed skins were in common
use in Egypt from the remotest antiquity; and the
Israelites carried numbers of them with them from
that country. "Rams' skins dyed red" seem to
have been held in high esteem by the Hebrews.
The under covering of the ark of the testimony
was formed of them. It is stated that every man
with whom was found rams' skins dyed red offered
them, and Moses made a covering for the tent of
rams' skins dyed red, and a covering of badgers'
skins above that.

C. Hamilton Smith, in Kitto's "Cyclopædia," says,
"We agree with Dr Mason Harris, that the skins in
question were most likely tanned and coloured crimson, for it is well known that what is now termed red
morocco was manufactured in the remotest ages in
Lybia, especially about the Tritonian Lake, where
the original *ægis* or goatskin breastplate of Jupiter
and Minerva was dyed bright red; and the Egyptians had most certainly red leather in use, for their
antique paintings show harness-makers cutting it
into slips for the collars of horses and furniture."
And Wilkinson says of ancient Egypt that they
made shoes, sandals, coverings and seats of chairs
and sofas; bow cases, and most of the ornamental
furniture of chariots, harps, and also shields were

adorned with coloured leather, and skins prepared in various ways requiring the dyer's art.

Dr Thomson in his "Land and the Book," says, "Salim led me through an entire street of shoe shops this morning. Is the red leather which the shoemakers use the rams' skins dyed red, which formed one of the three colours of the tabernacle? No doubt; and there is a definiteness in the rams' skins which is worth noticing. From time out of mind the southern part of Syria and Palestine has been supplied with mutton from the great plains and deserts on the north-east and south, and the shepherds do not ordinarily bring the females to market. The vast flocks which annually come from Armenia and Northern Syria are nearly all males; the leather, therefore, is literally rams' skins dyed red."

Wool.—That woollen stuffs were used, and spun and woven into cloth for common use, at a very early period of man's history, is beyond doubt; and that woollen was dyed various colours, either before or after being converted into cloth, is also plainly stated in history, sacred and profane. It is the opinion of Wilkinson that the ancient Egyptians dyed their woollen stuff in the state of wool or thread.

The words translated purple, blue, and scarlet, in Scripture, we are told, refer in the original to more than colour, implying rather the coloured material, such as purple, or blue cloth, or thread ; so that such passages as the following : " And this is the offering which ye shall take of them ; blue, and purple, and scarlet," embrace the materials dyed in these colours. Although woollen was in common use amongst the ancient Hebrews, it was held in very low estimation by the Egyptians, and was not worn by the better class of Egyptian society. No woollen was used in wrapping up their dead. The reason for this, according to Mr Wilkinson, was the preservation of the embalmed corpse, woollen cloth being liable to breed insects, which would destroy the body. This appears a satisfactory reason for the rejection of woollen fabric in this particular case, but it does not account for the general aversion to woollen stuff, which, we have said, prevailed among the richer classes of Egyptian society. Some writers, in order to account for the general aversion, consider it to have been an outcome of the hatred and loathing with which the shepherd class were regarded. Moses says, " Every shepherd is an abomination to the Egyptians." The Hebrew priests, while officiating in the tabernacle or temple, were forbidden to wear woollen garments as a matter of cleanliness. It is said, " They shall be clothed with linen garments ; and no wool shall come upon them while they minister ; they shall not gird themselves with anything that causeth sweat." Wool was extensively used by the Hebrews, as well as other nations, for carpets, which were dyed of rich colours.

SILK.—The existence of silk in ancient times has been doubted by several scholars. These have endeavoured to prove that the word translated *silk* in the Proverbs and Ezekiel, may with equal propriety be translated *fine linen* or *cotton*. It has been argued, from its great cost in the early period of our era, that silk could not have been known in ancient times. Aurelian complained that a pound of silk was sold at Rome for twelve ounces of gold, or weight for weight. But the same argument may be used with respect to the Tyrian purple. One pound wool dyed of that colour, in the time of Augustus, cost upwards of £32 of our money, hence it might be said that Tyrian purple could not have been known long before the age of Augustus. Under the article " Silk," in the " Encyclopædia Metropolitana," it is stated that silk was known in very ancient times.

Bonomi, in his book on " Assyria," says, " The custom of presenting robes as a mark of honour may be traced to the remotest antiquity in eastern countries, and even still prevails. The Midian habit was made of silk, and among the elder Greeks it was only another name for a silken robe. The silken robes of Assyria, the produce chiefly of the looms of Babylon, were renowned long after the fall of the Assyrian Empire, and retained their hold of the market even to the time of the Roman supremacy. Pliny speaks of the wool of silk, and believed it was washed down from trees. At a later period the western nations became acquainted with the worm or fly by which the silk is spun. As the worms were bred upon the mulberry trees, this may account for Pliny supposing that the wool of silk was washed down from trees. Silk was brought to Rome from Seres, a nation of Asia, supposed to be the modern Thibet, and was consequently called Sericum. Some have supposed that the Seres are the same as the Chinese, and reason from this that silk was known to them in the earliest ages. About two centuries before the Christian era silk was worth its weight in gold. In Rome it afterwards became very cheap, and was a common article of dress among the Romans, but it may have been known to other nations, previous to its being brought to Rome. Although the precise time that silk was introduced as a fabric for clothing in the more western countries cannot be fixed, still we are strongly of the opinion that the knowledge and use of it extend considerably into ancient times, at least to the days of Ezekiel, 594 B.C., who says, " Thus wast thou decked with gold and silver; and thy raiment was of fine linen, and silk, and broidered work."

Professor Rawlinson seems to have no doubt on this matter; speaking of the dress of the kings of Persia in times of peace, he says, " It was the long flowing Midian garment, or candys, made in all probability of richest silk."

COTTON.—Cotton was known to the ancients very early, but to what extent it was used in the earlier periods of man's history is not known. Herodotus is the earliest historian who mentions it, calling it a tree wool, observing that the trees of India bear fleeces as their fruit, surpassing those of sheep in beauty, and that the Indians wear clothes made from those trees. Ctesias, a contemporary of Herodotus, and who resided at the Persian court, speaks also of the Indian wool of trees. According to Baines, Yates, and several Biblical commentators, the word translated "green" in Esther i. 6, means "cotton," which shows that this fabric was used in Persia at least 100 years before Herodotus wrote; and it is not improbable that Solomon, who traded extensively with India, may have received cotton as part of his merchandise. Dr Royle tells us, in his "History of the use of cotton in India," that its manufactured products, and even the process of starching it, are referred to in the "Institutes of Menu" at as early a period as 800 years before Christ, close to Solomon's time. Egypt had also a trade with India at a much earlier time than that of Solomon, so that cotton may have been an article of commerce obtained from India as a matter of exchange. The mummy cloths of Egypt have been examined for proof of the use of cotton amongst that ancient people. But, contrary to expectation, all those examined by Bauer, Thomson, and Ure, were found to be composed of linen. Since these examinations, however, the late Dr Bowring ascertained that the mummy cloth of a child was formed of cotton, and not of linen as is the case with adult mummies. Rosselini has found the seeds of the cotton plant in a vessel in the tombs of Egypt, which is an indication that cotton was not only purchased as an article of import, but was cultivated in Egypt. Wilkinson says that cotton cloth was manufactured in Egypt, and that cotton dresses were worn by all classes. Pliny says that the Egyptian priests, though they used linen, were particularly partial to cotton robes; and on the Rosetta stone, there is mention made of cotton garments, supplied by the government for the use of the temple. Bonomi says, " The cotton manufactures were celebrated and remarkable, and are mentioned by Pliny as the invention of Semiramis, who is stated by many writers of antiquity to have founded large weaving establishments along the banks of the Tigris and Euphrates." The textile fabrics of Assyria were celebrated all over the civilised world. This implies an extensive shipping trade, and agrees with the words of Isaiah, 800 years before Christ, in reference to Assyria, " Thus saith the Lord, your Redeemer, the Holy One of Israel; For your sake I have sent to Babylon, and have brought down all their nobles, and the Chaldeans, whose cry is in the ships." From these evidences we think it proven that cotton was not only known, but used, in Egypt, Palestine, and all the other civilised countries in the East, from the earliest times of which we have any record or relic.

LINEN AND FLAX.—Whatever doubts may exist respecting the use of cotton and silk among ancient Eastern nations, there are none concerning their common use of linen. It is familiarly referred to in almost all ancient writings. The Old Testament speaks of linen, fine linen, and fine twined linen, &c. These being translations of different words in the original have given rise to some difficulties concerning the precise meanings of each term, some supposing that cotton may be the true meaning of some of these distinctive names ; however, that different qualities of linen were manufactured in ancient times, is amply verified by the still existent varieties found in the wrappings of mummies. Some of these cloths are so fine in texture that even at the present day we cannot surpass them, the quality of one piece found at Memphis " being," writes Wilkinson, " to the touch comparable to silk, and not inferior in texture to our finest cambrics." Coloured linens are also extant, of which we will speak hereafter. Linen was common amongst the Hebrews, as every reader of his Bible well knows. In Greece, and all the other nations of antiquity, linen was a common article of clothing; although, from a passage in Professor Ramsay's "Roman Antiquities," we may infer that linen as a whole was scarce, or certain qualities of it were expensive, "insomuch," says Ramsay, "that the priests of Isis were at once marked out to the eye as a distinct class by the circumstance of being robed in linen."

In our opinion all the various fabrics named were used less or more from very early times; some from the most remote antiquity, others from at least 700 to 1000 years before the Christian era. That they were all subjected to dyeing processes cannot be affirmed, but only considered probable. As these various fabrics are not acted upon alike by the same dye, the different kinds often requiring not only different treatment but different dyeing agents to produce the same effects, our ability to prove that one of these kinds of fabrics was dyed will be no evidence that the others were also dyed. We know, speaking from modern experience, that a dyer may excel in dyeing woollen, and be altogether unable to dye linen. Or, again, a dyeing agent may produce a beautiful red or crimson on woollen, and be quite unfit to dye linen; and these circumstances may be reversed, so that different treatment, and in many instances different dye-drugs, have to be used for animal and vegetable fabrics, so that in this inquiry these peculiarities will have to be borne in mind, as the ancients, like the moderns, would have to contend with these different properties. These matters we will be able to consider better after briefly considering what dyeing agents, so far as we know, they had to dye with. The following is the list:—

COLOURING MATTERS.—Kermes, indigo and woad, madder, archil, safflower, alkanet, henna, broom, galls, berries, walnut, pomegranate seeds, Egyptian acacia, and shell-fish.

SALTS.—Sulphate of iron (copperas), sulphate of copper (bluestone), acetate of copper (verdigris), acetate of iron (iron liquor), alum, alkaline carbonates, lime, and soap.

The possession of such articles as are named indicates a considerable knowledge of practical chemistry, although they may not have known anything of chemistry as a science.

A modern dyer confined to this list would find it very difficult to produce to satisfaction all the colours the ancients possessed, but it would be wrong to conclude that the ancients had not the ability to dye excellent colours with the materials named. Unfortunately our information about the application of these agents, and even about the agents themselves, is very meagre. The principal source of information is from general historians, the learned men of that day. Now, such arts as dyeing were held in very low estimation by these writers. Suppose that a general historian of the

present day, in one of his publications, gives
information concerning modern processes of dyeing
and the dyestuffs used, suppose also that he does
trouble himself to obtain correct information;
notwithstanding, we can safely affirm, by having
seen many examples, that his account would be
but a piece of popular writing, incorrect in many
particulars. This being the case, it is preposterous
to expect correctness from such ancient writers as
Pliny, who were too proud to visit workshops or
make inquiry of tradesmen concerning their processes. Pliny, on whom we mostly depend, evidently
had this mistaken pride; he says, " I should have
described the art of dyeing had it been included
among the number of the liberal arts." It is
therefore consistent with his own statement to
suppose that Pliny contented himself with the
floating knowledge wafted to him by popular report,
which concerning all arts is full of fallacies; besides,
Pliny lived at a time when the art of dyeing is
considered to have been much behind what it had
been centuries before. In Pliny's own time, at
all events of his own knowledge, he knew indigo
only as a pigment, and speaks of it as having been
got from the sea; hence it has been supposed by
some writers that indigo was not known as a dye
before his time. It may be that the Greeks and
Romans in Pliny's time could not dye with indigo;
but there is blue-dyed cloth still in existence,
dyed with indigo upwards of 1000 years before

Pliny was born. It is possible that this is one more proof of what we spoke of formerly of an art being lost and again rediscovered, or well known to one nation and not to another. We are inclined to think, however, that the art of dyeing with indigo was known to practical men in the days of Pliny, although not known to the historian. It is as likely as not that Pliny's silence arose from ignorance. It is known that such arts were confined to certain classes of the community, who kept their operations secret, so that much was known and practised by them which such men as Pliny could not possibly know. It is stated by Bischoff, "That in every province, and particularly in Phœnicia, there were certain houses for dyeing purple belonging to the emperors, and each of these was under the inspection of an overseer, whose chief business was to take care that the articles were well dyed. These overseers and their work were again under the inspection of a higher functionary." Probably it was one such general overseer that the king of Tyre sent to Solomon, "A man cunning to work in gold and in silver, and in brass and in iron, and in purple, and crimson, and blue." Neither the dyers nor their children durst follow any other occupation; they formed a peculiar tribe, and had their own symbol, which was a small basket containing purple and wool.

INDIGO AND WOAD.—The indigo-producing plants

have certainly been known from the remotest
antiquity.   This is attested by a variety of circumstances, particularly by its Hindoo name *Nil*,
which, according to Sir William Jones, means blue.
By the same name it was known to the Arabs, Egyptians, and other nations of the East, which is a
strongly circumstantial proof that indigo was first
imported from India.   But at the same time a
knowledge of the properties of indigo was possessed
by nations who had no intercourse with India, and
who, therefore, must have obtained their indigo
from native plants.   Cæsar states that the ancient
Britons made their blue with indigo from the
woad plant, and that their wives and daughters,
when they appeared naked at the sacred festivals,
had their bodies dyed with it, so that they resembled the Ethiopians.   Pliny conceived indigo to
be a slime naturally collected in the scum of the
sea, and adhering to certain reeds growing on its
shores, which Beckmann says was the indigo plant
stripped of its bark.   The peculiar smell which
indigo gives when burning, Pliny considered as
confirmatory of its marine origin.   When the
indigo dye or plant was first introduced into
Egypt, Palestine, and these localities, is not
known; but sufficient for our present purpose is
the fact that indigo was known and used by dyers
in ancient times for dyeing blues; and that it was
used to a great extent is evidenced by the extensive cultivation of the indigo plant in those

countries, which cultivation is believed to have been carried on from very early times. It is still cultivated in small quantities. Dr Bowring says it grows wild in certain parts of Palestine. Volney says it grows with cultivation on Jordan's banks. Burckhardt states that indigo is a common product of the eastern coast of the Dead Sea, and excels that grown in Egypt; and dyeing blue with indigo is still a common occupation of the people of Safet. That indigo was used as a dye by the ancient Egyptians is proved by the analyses of the blue upon mummy cloths. Mr Thomson who examined these colours says, "Though I had no doubt the colouring matter of the blue stripes was indigo, I subjected the cloth to the following examination :—Boiled in water for some time the colour did not yield in the least, nor was it at all affected by soap or strong alkalies. Sulphuric acid, diluted so far as not to destroy the cloth, had no action on the colour. Chloride of lime gradually reduced, and at last destroyed it. Strong nitric acid dropped upon the blue turned it orange, and in the same instant destroyed it; these tests prove the colouring matter of the stripes to be indigo." Loftus found in the graves of ancient Chaldea that the corpse had been dressed with blue linen, in all likelihood dyed with indigo.

KERMES.—These are the dried bodies of insects, which feed upon the leaves of the prickly oak. The word is considered Arabic, and signifies a

little worm. This substance was known as a
dyeing agent in the East before the days of
Moses, and was used from time immemorial in India
for dyeing silk and woollen, and was also a dry
drug, known to Greek and Roman dyers. We
find the following remarks upon the use of kermes
in ancient times in Beckmann's " History of
Inventions." Beckmann derives his information
from Professor Tychsen, who is the writer to
whom he refers in the extract :—" *Tola* was the
ancient Phœnician name for kermes, used both
by the Hebrews and the Assyrians ; and the
Assyrian translators of Scripture use that name
in Isaiah i. 18, called in our translation scarlet.
The dye was known to the Egyptians in the time
of Moses, and the Israelites must have carried it
along with them from Egypt. The same writer
thinks that the Arabs received the name kermes
from Armenia and Persia, where it was indigenous
and long known, and that that name banished
the old name, *tola*, in the East, as the name
scarlet did in the West." The following is Pliny's
account of kermes :—" There grew upon the oak
in Africa, Sicily, &c., a small excrescence like a
bud, termed *cusculium ;* the Spaniards paid with
these grains half of their tribute to the Romans.
Those produced in Sicily were the worst ; those
from the neighbourhood of Emerita the best, and
they were employed for dyeing purple."

SHELL-FISH.—When this means of dyeing was

discovered is uncertain. There are various accounts professing to give a narrative of the
discovery, but they are all mixed with fables.
The common story runs thus :—A young man
walking along the sea shore with his sweet-
heart, their little dog caught at a *Purpura*, by
which his mouth became stained with purple,
which the young woman observing, expressed a
desire to have a dress of the same colour ; her
lover, anxious to gratify her desire, examined and
found the shell, and so discovered the dye.
Others have attributed the discovery to the Phœnician Hercules, who was a great navigator. He
communicated his discovery to the king, who
immediately began to wear purple, and from that
time purple became a regal badge. The date of
the discovery has been given differently as 1500
B.C., and 1250 B.C., but purple is mentioned as
having been common in Egypt, and amongst the
Israelites, prior to the earliest of these dates, and
purple garments were worn by the kings of Midian
as early as 1291 B.C.

The shell-fish used for dyeing purple are named
by Pliny the *Conchylium, Murex, Purpura,* and
Buccinum. He says the best *Purpuræ* found in
Asia were those found in the sea adjoining Tyre;
that the Tyrians, when they caught any of the
greater *Purpuræ,* took the fish out of their shells,
the better to extract the colouring matter, but
that they obtained it from the smaller by grinding

in mills ; that the fishermen endeavoured to take
the purple fish alive, killing it immediately with
a blow, because if allowed to die slowly, it ejected
its precious liquor with its life.　He states also
that the fish dies speedily if put into fresh water.
The method of catching them was this :—The
fishermen took a net with wide meshes, into
which as bait a few mussels were put.　These
immersed in the water, open their shells, which
the purple fish observing thrust in their tongues,
but immediately they perceive this they again shut
their shells, and in this manner the *Purpuræ* were
caught.　When the *Purpuræ* was caught, the white
vein or receptacle of the colour was taken out and
laid in salt for three days ; after which the matter
so extracted and salted was boiled slowly in leaden
vessels over a gentle fire, the workmen from time
to time skimming off the fleshy impurities.　This
process lasted ten days ; after which the liquor
was tried by dipping wool into it ; and if the
colour produced was defective, the boiling was
renewed.　Pliny says this liquor was generally
used with other dyes, by which a variety of tints
were produced.　Amongst the other dyes are
named kermes, archil, and alkanet root.

By whatever means the dye was produced, purple
coloured cloth was certainly in ancient times held
in high estimation, and was at a later period
appropriated to the service of religion, and as a
distinguishing garb for the highest civil and

military dignities, and ultimately became the emblem and symbol of majesty. Under some of the Roman emperors the wearing of purple by any person not of the imperial family was deemed treason, and punishable by death. The natural effect of such restrictions would be that the skill to produce the colour in perfection would gradually deteriorate, and eventually all knowledge of the process become lost. Many attempts have been made in modern times to recover the knowledge of how to produce this celebrated dye, with various success, but without results sufficient to warrant it being tried as a manufacture. The late Dr George Wilson considered that the *murixide* got from guano was the same as the ancient purple, but in practice it does not give either the permanency or the beauty history ascribes to the Tyrian purple, probably from want of the proper mode of applica- tion. Our purples, and some of the other colours from coal tar, will stand favourable comparison with any purple ever produced either in ancient or modern times, so far as brilliancy is concerned ; but they also have not the permanency ascribed to the Tyrian purple. Mr Johnson, in his "Intro- duction to Conchology," makes mention of the colour obtained from shell-fish. He says, "Several shell-fish have the power of secreting and throwing out a fluid when irritated. Some of these fluids are purple-coloured ; they have been occasionally collected and tested, and several eminent men have

R

considered that this fluid, so excreted, may have
formed a part or whole of the celebrated purple
dye of ancient times." With this view Johnson
disagrees. As these excretions are from the first
a purple colour, and liable to change by acids and
alkalis, and fade by exposure, he thinks they
formed no part of the Tyrian dye, for unchange-
ableness was one of its characters ; and Aristotle
and Pliny state expressly that the colour of that
fluid, on its first discharge from the animal, was
white. Such a coloured liquid can be procured, as
these authors say it was procured, from several uni-
valves belonging to the genera *murex* and *puṛpuræ*,
and Colonel Montagu furnishes us with a good account of how it may be procured from the *Purpura
lapillus*,—" The part containing the colouring matter is a slender longitudinal vein, just under the skin
on the back, behind the head, appearing whiter than
the rest of the animal. The fluid itself is of the
colour and consistence of thick cream. As soon as
it is exposed to the air it becomes of a bright yellow,
speedily turns to a pale green, and continues to
change imperceptibly until it assumes a bluish cast,
and then a purplish red. Without the influence
of the solar rays it will go through all these
changes in two or three hours ; but the process is
much accelerated by exposure to the sun. A
portion of the fluid, mixed with dilute vitriolic
acid, did not at first appear to have been sensibly
affected ; but by more intimately mixing it in the

sun, it became of a pale purple, or purplish red, without any of the intermediate changes. Several marks were now made on fine calico, in order to try if it was possible to discharge the colour by such chemical means as were at hand; and it was found that, after the colour was fixed at its last natural change, nitrous, no more than vitriolic, acid had any other effect than that of rather brightening it. *Aqua regia*, with or without solution of tin, and marine acid, produced no change, nor had fixed or volatile alkali any effect. It does not in the least give out its colour to alcohol, like cochineal, and the *succus* of the animal at *Turbo* (*Scalaria clathrus*), but it communicates its very disagreeable odour to it most copiously, so that opening the bottle has been more powerful in its effects upon the olefactory nerves than the effluvia of asafœtida, to which it may be compared. All the markings which had been alkalised and acidulated, together with those to which nothing had been applied, became, after washing in soap and water, of a uniform colour, rather brighter than before, and were fixed at a fine unchangeable crimson."

I have myself obtained a beautiful permanent purple from *Purpura lapillus* found in Arran. The experiments of Colonel Montagu are important, as they were made on cotton, and found fast without the addition of a mordant, excelling in this respect any other colouring matter we know, except indigo; even the tar colours require a mordant for fixing

them upon cotton. In "The Land and the Book,"
Dr Thomson, speaking of ancient Sidon, says, "The
Tell on which the castle stands is artificial, and
what is more remarkable, is made in great measure
of old pottery, rubbish of houses, and thick beds
of broken *Purpura*, thrown out from Sidon's ancient
manufacture of purple dye." And again, speaking
of Tyre, "A variety of the *murex*, from which
the purple dye was procured, is found all along
this coast, but it abounds most around the Bay of
Acre ; so also the *Helix Jauthina*, from which a
blue, with a delicate purple of lilac tinge, may be
extracted, is equally abundant. After a storm in
winter you may gather thousands of them from
the sandy beach south of Sidon. They are
so extremely fragile that the waves soon grind
them to dust. A kind of *Ruccinum* is found here
at Tyre, which has a dark crimson colouring
matter about it, with a bluish, livid tinge. According to ancient authors this was used to vary
the shades of the purple. Pliny says the Tyrians
ground the shell in mills to get at the dye. This
could not have been the only process, because the
remnants of these shells found in pits along the
south-eastern shore of the island 'were certainly
broken or mashed, and not ground, and the same
is true of the shells on the south of the wall at
Sidon." .

MADDER.—This important colouring matter was
known to the ancients as a dye-drug. Pliny says

expressly that the red roots of the *rubia* were used
to dye wool and leather red. It is also stated by
Beckmann, quoting Dioscorides, that the thin, long
roots, &c., which are red, serve for dyeing; and on
that account the cultivated kind (which indicates
an acquaintance with a wild sort) is reared with
much benefit in Galilee, around Ravenna in Italy,
and in Caria, where it is planted either among the
olive trees or in fields destined for that purpose.
Virgil refers to the madder plant, and says the
sheep feeding upon it had their wool coloured red.
It is known that swine and several other animals
have their bones coloured red by eating madder.
Dr Kitto, in his " Physical History of Palestine,"
says madder grows abundantly in Syria and
Palestine. The dye from the madder root was certainly in very common use among the Egyptians,
and doubtless also among the Hebrews. The
reddish coloured dye of the mummy cloths appears
to have been produced from madder, and remains
a curious and interesting monument of its use.

CARTHAMUS, or safflower, was, in very remote
times, cultivated in Egypt, where it is still cultivated, producing the best quality of the dye.
Wilkinson says it is now proved by the discovery
of the seeds of *Carthamus tinctorius* in a tomb in
Thebes that the plant was cultivated in ancient
Egypt, not only for the dye it produced, but for
an oil extracted from its seeds.

ARCHIL.—Theophrastus and several other ancient

writers mention a plant which grew upon the rocks
of different islands, particularly in Crete or Candia,
and which had been used for a long time as a
violet dye for wool, said by some to have excelled
the ancient purple.   This plant Beckmann and
several others consider to have been the lichen
Roccella, our archil.   Pliny says that with this
plant dyers gave the ground or first tint to those
cloths they intended to dye with the costly purple.   We think it very probable that this method
of bottoming the purple with archil was only introduced at a later date, probably after the true
purple dye became scarce and costly, and is
an early instance of the production of a cheap
substitute.   Archil dye is not permanent, and
although topped with the real purple dye, and
when fresh equally brilliant, yet the dye would
not remain permanent, the foundation of it being
bad.   Probably the prohibition against dyeing and
wearing the real imperial purple led to this spurious
substitute.

HENNA.—This plant, which is abundant in
Egypt, Arabia, and Palestine, was used by the
ancients, as it is by the moderns, for dyeing.   The
leaves were dried and pulverised, and then made
into a paste.   It is a powerful astringent dye,
and is applied to desiccate and dye the palms of
the hands and soles of the feet and nails of both,
and gives a sort of dun or rust colour to animal
tissues, which is very permanent.   It is stated that

when sal-ammoniac and lime were put upon the coloured parts they changed to a dark greenish-blue colour, and passed on to black, probably from the sal-ammoniac containing iron, which would give this result. The Tyrian ladies dyed rings and stars upon their persons. Men gave a black dye to the hair of their heads and beards. The dyeing of the nails with henna is a very ancient custom. Some of the old Egyptian mummies are so dyed. It is supposed that the Jewish women also followed this custom. Reference is made to it in Deuteronomy, where the newly-married wife is desired to stain her nails. Also, in the Song of Solomon, *Camphire*, in the authorised version, is said to mean henna, which· has finely-scented flowers growing in bunches, and the leaves of the plant are used by women to impart a reddish stain to their nails. Speaking of Arabian women at the present day, Dr Thomson, in " The Land and the Book," says, "They paint their cheeks, putting tahl around their eyes, arching their eyebrows with the same, and stain their hands and feet with henna thus to deck themselves, and should an unmarried woman do so, an impression is conveyed highly injurious to the girl's character." This beautifully illustrates the passage just referred to (Deut. xxi. 12).

GALLS are named ·among the substances known. to the ancients, but I cannot find whether they were used as a dyeing agent. Wilkinson says that

tanning was in Egypt a subdivision of dyeing,
and it is mentioned that copperas with galls dyed
leather black; and there can be little doubt that
galls were used for a similar purpose in ordinary
dyeing. The *Myrobollans* and several sorts of
barks and pods of the *Acacia nilotica* were also
used for tanning, from their astringent properties,
and may have been similarly used for dyeing.

These are a few of the principal colouring matters used by dyers in ancient times. There is a
little confusion with respect to some of the salts
mentioned as having been used by them, especially
the alkaline salts—a circumstance, however, not
to be wondered at. In more modern times there
is a similar confusion on this same head.

When nitre, for instance, is burned with carbonaceous matter, the product is carbonate of
potash. The ashes left by burning wood contain
the same salt. The ashes left by burning seaweed produce carbonate of soda. When nitre is
burned with sulphur, the product is sulphate of
potash, &c. These have all been called generically,
even in modern times, nitre, having each a certain
prefix well understood by the adept, or chemist, of
the day. We think it probable that all these
processes for making the different salts were
practised in ancient times, but now having only
the generic name *nitre* given us by historians, we
cannot understand exactly when nitre is mentioned
which of the nitres is meant. When Solomon

speaks of the action of vinegar upon nitre, the chemist understands that the salt referred to is a carbonate, but when the nature of the action or application is not given, we have no idea what particular salt is meant. There is no doubt, however, that the ancients were well acquainted with the alkaline salts of potash and soda, and applied them in the arts. The metallic salts of iron, copper, and alumina were well known, and their application to dyeing was generally the same as at the present day. That they were used both as mordants and alterants is evident from several references. A very suggestive statement is made by Pliny about the ancient Egyptians. "They began," says he, "by painting or drawing on white cloths with certain drugs, which in themselves possessed no colour, but had the property of attracting or absorbing colouring matter; after which these cloths were immersed in a heated dyeing liquor; and although they were colourless before, and although this dyeing liquor was of one equable and uniform colour, yet when taken out of it soon afterwards, the cloth was found to be wonderfully tinged of different colours, according to the peculiar nature of the several drugs which had been applied to their respective parts, and these colours could not be afterwards discharged by washing."

Herodotus states that certain people who lived near the Caspian Sea could, by means of leaves of trees which they bruised and steeped in water,

form on cloth the figures of animals, flowers, &c.,
which were as lasting as the cloth itself. This
statement is more suggestive than instructive.
Persia was much famed for dyeing at a very early
period, and dyeing is still held in great esteem in
that country. Persian dyers have chosen Christ
as their patron; and Bischoff says that they at
present call a dye-house Christ's workshop, from
a tradition they have that He was of that profession. They have a legend, probably founded
upon what Pliny tells of the Egyptian dyers,
" that Christ being put apprentice to a dyer, His
master desired Him to dye some pieces of cloth
of different colours; He put them all into a boiler,
and when the dyer took them out he was terribly
frightened on finding that each had its proper
colour." This or a similar legend occurs in the
apocryphal book entitled " The First Gospel of
the Infancy of Jesus Christ." The following is the
passage :—" On a certain day also, when the Lord
Jesus was playing with the boys, and running
about, He passed by a dyer's shop whose name
was Salem, and there were in his shop many
pieces of cloth belonging to the people of that
city, which they designed to dye of several colours.
Then the Lord Jesus, going into the dyer's shop,
took all the cloths and threw them into the furnace. When Salem came home and saw the cloth
spoiled, he began to make a great noise and to chide
the Lord Jesus, saying, ' What hast Thou done
unto me, O thou son of Mary ? Thou hast injured

both me and my neighbours; they all desired their cloths of a proper colour, but Thou hast come and spoiled them all.' The Lord Jesus replied, ' I will change the colour of every cloth to what colour thou desirest,' and then He presently began to take the cloths out of the furnace; and they were all dyed of those same colours which the dyer desired. And when the Jews saw this surprising miracle they praised God."

TIN.—We have no positive evidence as to whether the ancients used oxide, or the salts of tin, in their dyeing operations. A modern dyer could hardly produce permanent tints with some of the dye drugs named without tin salts. We know that the ancients used the oxides of tin for glazing pottery and painting; they may therefore have used salts of tin in their dyeing operations. However, they had another salt—sulphate of alumina —which produces similar results, although the moderns in most cases prefer tin, as it makes a more brilliant and permanent tint.

ALUM.—This is what is termed a double salt, and is composed of sulphate of alumina and sulphate of potash. The process of manufacturing it in this country is by subjecting clay slate containing iron pyrites to a calcination, when the sulphur with the iron is oxidised, becoming sulphuric acid, which, combining with the alumina of the clay, and also with the iron, becomes sulphate of alumina and iron; to this is added a salt of potash, which, combining with the sulphate of alumina, forms the double

salt alum. Soda or ammonia may be substituted
for potash with similar results; the alum is crystallised from the solution. That the ancients were
acquainted with this double salt has been disputed,
but we think there can be no doubt of its existence
and use at a very early period. A very pure alum
is produced in volcanic districts by the action of
sulphurous acid and oxygen on felspathic rocks,
and used by the ancients for different purposes.
Pliny mentions *Alumine*, which he describes as
white, and used for whitening wool, also for dyeing
wool of bright colours. Occasionally he confounds
this salt with a mixture of sulphate of alumina
and iron, which, in all probability, was alum
containing iron, the process of separation not
being perfect; and he mentions that this kind of
alumen blackens on the application of nut-galls,
showing that iron was in it. Pliny says of
alumen, that it is "understood to be a sort of
brine which exudes from the earth; of this, too,
there are several kinds. In Cyprus there is a
white alumen, and another kind of a darker colour;
the uses of these are very dissimilar, the white
liquid alumen being employed for dyeing a
whole bright colour, and the darker, on the other
hand, for giving wool a tawny or sombre tint."
This is very characteristic of a pure aluminous
mordant, and of one containing iron. He also
mentions that this dark alumen was used for
purifying gold. He must be referring here to its
quality of giving to gold a rich colour. The liquid

of this iron alumen, if put upon light-coloured gold, and heated over a fire, gives it a very rich tint; a process practised still for the same purpose. So far, however, as the application to dyeing is concerned, it is unnecessary to prove that the ancients used our double salt alum. Probably the alumen referred to by Pliny, as exuding from the earth, was sulphate of alumina, without potash or soda, a salt not easily crystallised, but as effective, in many cases more effective, in the operations of dyeing than alum, which is attested by the preference given to this salt over alum for many purposes at the present day. Pliny says that alumen was a product of Spain, Egypt, Armenia, Macedonia, Pontus, Africa, and the Islands of Sardinia, Melos, Lipara, and Strangyle, and that the most esteemed is that of Egypt. And Herodotus mentions that King Amasis of Egypt sent the people of Delphi a thousand talents of this substance, as his contribution towards the rebuilding of their temple. Notwithstanding considerable confusion in Pliny's account of this substance, our belief is, that it refers to different salts of alumina, and whether or not they were all used in the processes of dyeing, they were used for manufacturing purposes, and thus gives us some insight to the advanced state of the arts in these times.

SOAP.—We take the following from the pen of F. T. Buckland :—" It has been lately suggested by an eminent lecturer that the ancient Greeks and Romans were not acquainted with the use of this valuable material; and that it is very possible that the flowing robes and graceful drapery which look so pretty when cut in marble, were not of the cleanest kind. The following passage from 'Knight's Pictorial Bible' throws some light on this point: Mal iii. 2, 'For he is like a refiner's fire, and like fullers' sope.' 'Fullers' sope.' The word 'soap,' by which the Hebrew 'borith' is translated, might lead the general reader to suppose the Hebrews possessed such soap as is in use among ourselves. Such was not the case. The word 'borith' is translated by the Septuagint, followed by the Vulgate, 'Fullers' herb,' whence, and from the explanation of the rabbis, as well as from our knowledge of the substances anciently, and even now, employed in the place of soap, we may collect that the purifying substance was a vegetable alkali, obtained from the ashes of an alkaline plant. This was used, or a solution of it, in connection with oil, for washing clothes in ancient times, and continues to be employed for the same purpose in different parts of the East. As there are several plants which furnish the requisite alkali, it is doubtful what particular plant, or whether any one alkaline plant in particular, may be intended. The substance may have been obtained from different plants, and it appears to us that the

name 'borith' denotes not the plant which
furnished the substance, but the substance itself,
from whatever plant obtained. Jerome, however,
supposes that the substance was furnished by a
particular plant growing in Palestine in moist and
green places, and which had the same virtue as nitre
to take away filth. Maimonides says the plant was
called gazul in the Arabic language. Although this
borith be that which our version renders ' soap,'
we are not to suppose that the Hebrews employed
no other substance for purification. The Bible
itself (Prov. xxv. 20, Jer. ii. 22) mentions a mineral
alkali (nitre) as employed for the same purpose;
and the Misnah counts the ' borith ' but as one of
seven things employed to extract spots and dirt
from clothing."

The substance referred to in Proverbs is no
doubt carbonate of soda, and the same is most
probably that referred to in Jeremiah. H. B.
Tristram says that on all sides of the south end
of the Dead Sea, the soap plant (*Salicomia
fructicosa*) and *Salsolas* abound, and (alluded to in
Jeremiah and Malachi) the Bedouins burn it for
alkali to use as soap, just as kelp used to be
burned on the Scottish coast. This would produce
carbonate of soda, with strong alkaline property.
Dr Thomson, in " The Land and the Book," mentions that great mounds of grey ashes of the soap
manufacture exist in the East, and that such mounds
are found near Jerusalem and Gaza, and adds, " I

cannot account for these immense hills of ashes, except on the supposition that kali, used in the manufacture of soap, has been very impure, leaving a large residuum to be cast out in heaps." " In Syria it is obtained mostly from the Arabs of the frontier deserts, where it is made by burning the *glass wort* and other saliferous plants that grow in these arid plains. The kali resembles in appearance cakes of coarse salt, and it is generally adulterated with sand, earth, and ashes, which makes the residuum very large, and from it these vast hills of rubbish gradually accumulate round the place where soap is manufactured. The growth of these mounds, however, is so slow, that it must have taken hundreds, if not thousands of years, for those of Edlip to reach their present enormous size." From these remarks we think it pretty evident, that soap, even in our meaning of the term, was known to the ancients. Within these sixty years most of the soap made in this country was obtained from the ashes of seaweed.

DIFFERENT COLOURS.—From a consideration of the list of dye stuffs and salts used as mordants— the whole of which I think we have not got an account of—the ancients may have been able to produce a considerable variety of colours and tints ; but, as might have been anticipated, there are very few colours named, owing, no doubt, to a confusion in the naming of tints. In Scripture we have only blue, red, crimson, purple, and

scarlet; but what distinguished red from crimson
or scarlet we do not know, but will take them as
translated, with the present definitions. Profane
history mentions, in addition to these, yellow,
green, and black; but they could not fail to have all
the intervening tints. It is worthy of remark that
in Scripture there is no reference to either yellow or
green. The Hebrew people being so fond of gold
ornaments, we would naturally reason that the colour
of gold would be a prevailing tint in their dress;
but colours, as most things else in the East, were
symbolical. Yellow was a symbol of subjection,
and although esteeméd, yet, according to Pliny, it
was exclusively worn by women; so that, if the
same idea existed in early times, the Hebrews
would avoid it, especially in connection with their
religious services. It is stated that the veils
which brides wore on their wedding-day were
yellow—a symbol of promise to serve, honour, and
obey. The dyeing of the nails yellow, as formerly
noticed, may have had a similar meaning. One
reason may account for the absence of yellow dye
in their embroidery for the tabernacle is their
lack of a permanent yellow dye. Linen dyed
yellow by any of the dye drugs which we know
they used, was fugitive, passing speedily into a
very dirty faded tint, which would have destroyed
the effect of the whole embroidered figure. That
this had something to do with its exclusion is
rendered probable by the use of gold thread along

s

with the other dyed threads in their embroidered
work. "And they did beat the gold into thin
plates, and cut it into wires, to work it in the
blue, and in the purple, and in the scarlet, and in
the fine linen, with cunning work." Wilkinson,
speaking of the combination of colours in Egypt,
says, that when black was used, yellow was added
to harmonise with it. A yellow colour on some
mummy cloth, being tested by Mr Thomson, gave
no trace of tannin, and seems to have been dyed
by some extractive matter; however, so far as we
have seen, the so-called yellow upon mummy cloth
is more of a dun or straw colour, and would not
pass for a yellow by either dyer or painter of the
present day. We must not forget, however, that
yellow is a favourite colour with the Chinese;
indeed, it is their royal colour, and has been so for
ages. The ancient Persians also held yellow in
high esteem, probably they had a proper yellow
dye. The symbolism of the colour also may have
been different among the Persians and Chinese,
from what it was among the Egyptians and
Hebrews.

SCARLET.—This colour was in the East a symbol
of triumph and rejoicing, and is still used as such
in India. Scarlet flags are used on the temples,
and as personal exhibitions, and indicate security
and joy. During the feast of the Hooli the inhabitants are in the habit of scattering cinnabar;
and on all great Hindoo festivals it is the custom

to wear necklaces of scarlet silk or worsted thread, so that in the East a scarlet thread or cord is a thing easily procurable in any house, and ready at hand. We find the first mention of it in Scripture, several years before the family of Jacob went into Egypt (Gen. xxxviii. 28). Indeed, it is the first dye mentioned in history, and although only incidentally, yet in such a common and familiar way as to show that the colour was well known. Again, on another and interesting occasion, a scarlet cord was easily procured by Rahab; and in this instance it was not without its symbolical meaning, of a token or covenant of security. In some of the Hebrew rites, such as the ceremony of pronouncing a person or house free of leprosy, scarlet was used along with hyssop, which may have been a symbolical emblem of joy and triumph; and the spreading of a scarlet cloth over the ark of the covenant, during its movement from place to place, may also have been a symbolical expression of security. Scarlet is also mentioned as the dress of females in David's lament for Saul, and probably in its symbolic sense—

" Ye daughters of Israel, weep over Saul,
    Who clothed you in scarlet and delightful garments,
    Who put on ornaments of gold upon your apparel."

Solomon represents the virtuous woman as providing scarlet clothing for her household as a protection against the cold. Thus " she is not afraid of the snow for the household, for all her household are clothed in scarlet." In China scarlet and bright red are the festive colours, denoting triumph, success, or joy; scarlet or red painted sign-boards are hung out on festive occasions. In Babylon, Daniel is said to have been clothed in scarlet by the king, but in this case it is considered by scholars that purple is meant. It is impossible to tell whether the word translated scarlet in Scripture refers to the same tint we designate by that name; but from the fact already stated, that it refers to kermes, it is evident that if not that fiery red we call scarlet, it was a bright red quite distinct from crimson and purple. We have no evidence as to the fabric on which this colour was dyed, but if it was on linen or cotton, and the dye kermes, then it is one of the lost arts. However, they might have dyed a bright red from safflower upon linen; but this is a fugitive dye. They might also obtain a red colour with madder; but neither of these gives the tint of kermes. It is more than probable that scarlet was generally dyed upon worsted; but even upon that fabric we could not dye a very good or permanent colour with kermes without a tin mordant. It is very probable they had tin dissolved as a mordant, although there is no evidence to prove it; but whatever was the mordant used, the dye appears to have been fast. We have a beautiful figure drawn from its permanence by the prophet Isaiah —"Though your sins be as scarlet, they shall be as white as snow; though they be red like crimson, they shall be as wool." Had the prophet in this poetic symbol been intent upon the extreme opposite of purity, he would have named black; but this colour is not permanent on any fabric, and is easily removed; on this account it would not have suited the lesson meant to be conveyed. This was not the case, however, with scarlet. The word which has been translated crimson here, is universally admitted by scholars to be wrong, and should be translated " double scarlet," or "scarlet double dyed;" and the passage should read— " Though your sins be as scarlet, they shall be as white as snow; though they be red as scarlet double dyed, they shall be as wool." A colour so dyed upon wool could not be removed without destroying the cloth. Scarlet is referred to by Nahum, who wrote nearly 800 years before Christ, as the dress of soldiers—

"The shield of his mighty men is made red;
The valiant men are in scarlet;"

doubtless as a symbol of triumph.

RED.—The word red is mentioned in Scripture repeatedly in connection with scarlet—red as scarlet, &c.; but as a distinct tint or dye it is only mentioned in Scripture in connection with rams' skins, which were used as coverings for the furniture of the tabernacle. Some commentators think this referred to the skins of a certain kind of sheep, having wool of a reddish-brown or fawn colour, much esteemed in the East; but in all the passages it is distinctly stated rams' skins *dyed* red, except in one place, where it is red skins of rams, so that I think we are not warranted in supposing they were natural, but were dyed in the tanning. Coloured leather, as before-mentioned, was common in Egypt long before the exodus. Red, in Egypt, was emblematic of the earth or earthy; the same idea was common among the Hebrews, and appears in the name Adam, red earth, and in Edom, which is red.

CRIMSON.—This is another colour which is very seldom referred to as a particular dye. It is mentioned as one of the three colours of the vail in Solomon's temple. "And he made the vail of blue and purple and crimson, and fine linen, and wrought cherubims thereon." It is stated of Hiram that he was cunning to work in purple and crimson, and blue, by which is meant, evidently, a knowledge of the artistic combinations of the colours used in embroidery.

Crimson is once mentioned in Jeremiah; but here, however, it is universally agreed that the translation should be scarlet, the reference is evidently to the symbolical meaning of scarlet. "And when thou art spoiled what wilt thou do? Though thou clothest thyself with crimson (scarlet), thy lovers will despise thee, they will seek thy life."

BLUE.—As a dye this colour is referred to very

early, and as might naturally be expected had a
high symbolical application. Blue, the colour of
the sky—the seat of the gods—became a colour
emblematic of heaven, but it soon begot a superstitious belief. There is a beautiful poetic illustration of this symbolic meaning in a picture found
in a mummy-case a short time ago. The blue
vault of heaven is represented by the goddess
Ne'ith. Beneath this is a double figure (two
bodies) representing the deceased; his earthy body
is red and in the act of falling to the ground,
whilst his spiritual or heavenly body is blue,
standing upright and raising his hands to heaven.
In this symbolical picture is graphically shown
the Egyptian belief in the immortality of the soul.
In the East, blue was the colour of protection,
and continues so till the present day. It may
have been in consequence of this symbolism that
blue was the colour selected for the cloth covering of
the tabernacle. "In the East," says Mrs Postans,
"they wear blue as a protection against an evil
eye." "I inquired," says she, "why the favourite
mares of the Balooche chiefs had necklaces of blue
beads, and was told it was to protect them from
an evil eye. My water-drawer always saw that
the one blue ball was securely tied round the
throat of his little bullock; and a Hummall in
my service in India, who had been a sufferer from
a stroke of the land-wind, at once tied a blue
cotton thread round his ankle, on which, he said,

the evil spirit that tormented him would be
obliged to fly.　The turquoise stone is often worn,
in consequence of its colour, as a protection for
the wearer against disease and evil eye."　It is
possible that some of these ideas and supersti-
tions had their origin in higher sentiments.
Moses is commanded to direct the children of
Israel to put upon the fringe of the borders of
their garments a ribbon of blue, as a memorial,
which, when they looked on it, might remember
them of their God and His commandments.　In the
transit of the tabernacle furniture, during their
march in the wilderness, all the sacred things,
the ark, candlestick, &c., were covered with blue
cloth.　The curtains of the tabernacle were
attached to each other by loops of blue, fifty
loops to each curtain ; the breastplate was bound
to the girdle by a lace of blue.　The plate of gold,
whereon was written "HOLINESS TO THE LORD," was
tied by a lace of blue to the front of the mitre.
The robe of the ephod worn by the high priest
was all of blue.　Indeed, this colour seems to
have been what may be termed the covenant
colour of the Hebrews.　In connection with the
tabernacle, the blue, purple, and scarlet, combined
in embroidered work upon fine linen, may have
held their symbolic meanings—purple, regality ;
blue, protection ; scarlet, triumph ; and white of
the linen, purity.　A combination of the same
colours formed the vail of Solomon's temple.　In

Babylon, blue, it would appear, was a favourite colour worn by the highest classes, whether because of any superstitious idea of protection attached to it is not stated. Ezekiel writes, "She doated on her lovers, on the Assyrians her neighbours, who were clothed with blue, captains and rulers." Loftus tells us, in his "Researches in Ancient Chaldea," that in ancient burying-grounds, the skeleton is found holding cylinders of agate with a copper bowl, and that the corpses have been dressed with blue linen.

These facts show the prominence and common use of blue dye in ancient times; and there is little doubt the dye was indigo, not only from the fact that indigo was in possession of the ancients, but that blue linen still extant is found to be dyed with indigo. Whether they dissolved their indigo by the same means as we do is not known; but they had the same reagents wherewith to do so, for the dyeing of both animal and vegetable fabrics. Though the Greeks at the time of Pliny may not have known indigo as a dye, still it is certain that other localities produced the dye from indigo at that time, as well as for nearly two thousand years previously. Certain localities where the inhabitants devoted themselves to certain particular branches of trade, became, as was natural, famous for their speciality. Ezekiel, who gives a list of some of their localities and their specialities in reference to dyed colours, has the following passage—

> "Fine linen, with broidered work from Egypt,
> Was that which thou spreadest forth to be thy sail ;
> Blue and purple from the Isles of Elisha
> Was that which covered thee."

PURPLE.—According to popular belief this is the colour *par excellence* of the ancients, and the encomiums passed upon this colour by ancient authors have given to the art of dyeing a position in ancient history, which, in the popular mind, it otherwise would not have held. The Tyrian purple is described as surpassing in brilliancy every other dye, although no doubt some of our coal tar colours are equally beautiful. Purple, like the other colours, had an emblematic significance. Whether it had a religious symbolism I have not been able to ascertain, but close upon the commencement of the Christian era we find it an emblem or badge of royalty, or royal favour, particularly in Rome, where those who had the right to the distinguished honour of wearing purple, were defined and decreed by a legal enactment. We think commentators on Scripture, and some other writers, have misapplied this circumstance, by considering that it applied to all ancient times and nationalities, and have too hastily identified all purple colours of which they find mention with the Tyrian purple dyed by shell-fish. Now, the Egyptians, Hebrews, and, we believe, other nations, had purple cloth and thread long before even the earliest date ascribed to the discovery of the Tyrian dye.

In fitting up the tabernacle, where a leading feature was to impress on the Hebrews to regard God as an earthly as well as heavenly king, purple is less prominent than blue and scarlet, although abundant, and this was prior to the discovery of the Tyrian purple. The only place where purple is named and used apart from other colours is as follows :—"And they shall take away the ashes from the altar, and spread a purple cloth thereon: and they shall put upon it all the vessels thereof, wherewith they minister about it, even the censers, the flesh-hooks, and the shovels, and the basons, all the vessels of the altar." These were afterwards covered with skins, and so carried. It is difficult to see any special emblem of royalty in this use of purple cloth. The first mention of the use of purple by royalty is in the time of Gideon, between two and three hundred years after the erection of the tabernacle, when, amongst the spoils taken from the enemy, is named the purple raiment that was on the kings of Midian. In the furniture of the temple of Solomon purple gets no higher prominence than the other colours used along with it for tapestry and embroidered work. However, Solomon, in his allegorical song, mentions purple as the covering for the royal

chariot. The same writer speaks of it as a dress
worn by an industrious and frugal wife—

"She worketh beautiful vestments for herself;
Her clothing is fine linen and purple."

That the wearing of purple was held as a mark of
high honour in Babylon, in the times of Jeremiah,
is evident from the fact that their gods are spoken
of as being clothed in purple and blue, these
colours being probably emblematic of power and
protection. In Persia, in the time of Esther,
about 500 or 600 years before Pliny, Mordecai is
ordered to be clothed in royal apparel, which is
stated to be blue and white, with a garment of
fine linen and purple. It is probable that the
adoption of purple robes or garments as a badge of
royalty or nobility was gradual, and limited to
certain nations; and that not till the palmy days
of Rome was it limited to royalty, or distinctive of
royal favour, although royalty and the higher
classes had their marks of distinction in dress in
all ages and nations, and purple may have formed
a part, but only a part, of such dresses. It is
well known that dresses of divers colours have
been marks of distinction from the earliest periods,
and are still so in the East. Jacob gave Joseph a
coat of many colours, the symbols of which his
brethren well understood, and it increased their
enmity. Some suppose this coat was like our
tartan, others that it was made of patches of

different coloured cloth. Roberts, in his "Illustrations of the East," says, "For beautiful and favourite children, crimson and purple and other colours are tastefully sewed together. Sometimes children of Mohammedans have their jackets embroidered with gold and silk of various colours; a child being clothed in a garment of many colours, it is believed that neither tongues nor evil spirits will injure him." In Persia, India, and China, vestures composed of various colours are still worn as marks of distinction. Embroidered garments of various colours have been emblems of the highest distinction, even of royalty, from the most ancient times. Such was the distinctive badge of the furniture of the tabernacle, and such was the highest mark of honour an ambitious and loving mother is represented as seeking for her son—

> "Have they not sped ! have they not divided the spoil ;
> To Sisera a spoil of divers colours,
> A spoil of divers colours of needlework,
> Of divers colours of needlework on both sides,
> Meet for the necks of those who take the spoil !"

Homer indicates this embroidered work as a dress of kings, thus—

> "The bitter weapon plunged into his belt,
> Transpierced the 'broidered cincture through its folds,
> His gorgeous corselet," &c.

He also refers to Tyrian dye, thus—

> "Belts
> That rich with Tyrian dye refulgent glowed."

This is presumed to refer to purple, which is called the Tyrian dye.

In David's time garments of divers colours were worn by the king's daughters before they were married. Pliny refers to embroidered and also woven cloths of different colours as being known to Homer, and states that the Babylonians were the most noted for their skill in this kind of work. It was, no doubt, one of these wrought and coloured garments that tempted Achan to break Joshua's order. In his confession he says, "When I saw among the spoils a goodly Babylonish garment, &c., I coveted them, and took them."

Professor Rawlinson in his "Five Ancient Monarchies," says, in reference to the dresses figured on the ancient seal of the king of Urr, the successor of Nimrod the mighty hunter, "We may conclude without much danger of mistake, that in the time of the monarch who owned the seal, dresses of delicate fabric, and elaborate pattern, and furniture of a *recherché* and elegant shape, were in use amongst the people over whom he exercised dominion. The figures on the above seal have flounces and fringes delicately striped, which indicates an advanced state of this kind of manufacture five or six centuries before the time of Joshua."

The Egyptians also were skilled in this sort of work. Wilkinson says, "Many of the Egyptian

stuffs presented various patterns worked in colours
by the loom, independent of those produced by
the dyeing and printing process, and so richly
composed that Martial says they vied with the
Babylonian cloths embroidered with the needle.
The weaving of patterns of different colours into
cloth is very suggestive of high attainments in that
art." From these, and many other references, we
are of opinion that although the purple colour was
symbolical of power and authority, it was not
worn alone in the more early times, but combined with other colours, and that neither was
this colour itself exclusively worn by people in
power or authority, but was worn in many ways,
and put to various uses, by those who could afford
it. The dye produced from shell-fish seems to
have been very costly, which naturally made it
exclusive. The sails of vessels, in the earliest
times, were made of rich colours, with fanciful
devices embroidered upon them, some representing
the soul of the king, flowers, and other patterns;
while others were adorned with coloured checks or
stripes. Embroidered sails were long a manufacture
of Egypt, and seem to have been bought by the
Tyrians for that purpose, as stated by Ezekiel—
" Fine linen, with embroidered work from Egypt,
was that which thou spreadest forth to be thy
sails." These sails, Wilkinson says, were used
at a very early period in Egypt; and the hem or
border of the sails was neatly coloured. The use

of these sails was mostly confined to the pleasure-
boats of the grandees and of the king, ordinary
sails being white.   It is stated that the devices
and colours on these sails denoted the rank of the
party.   Such a custom as this, doubtless, gradually brought into use certain colours or combination of colours as symbols distinctive of rank.
And as wealth always imitates the symbol of rank,
an exclusive right to certain rank marks and
colours might come to be enforced by kings, in
order to protect their own symbols.   It is mentioned by Atticus that the sails of the large ship of
Ptolemy Philopater were of fine linen, ornamented
with a purple border.   Upwards of a century after
this, Julius Cæsar prohibited the use of purple to
his subjects, except upon certain days.   After this
time we find the ship of Antony and Cleopatra at
the battle of Actium, distinguished from the rest
of the fleet by having purple sails—a distinction
which is said to have been at that time the peculiar privilege of the admiral's vessel.   In the
time of Nero the wearing of purple without his
authority was punished with death—a restrictive
law, which, I think, indicates that previously the
use of the colour was common, although that particular sort of it known as Tyrian purple may, from
its very great cost, have been confined to the
wealthy.   This restriction could scarcely be against
the use of all purple dyed cloths, or if so, it must
have remained in force for only a short time,

otherwise, it is irreconcilable with other notices of
about the same period.   In the list of imports and
exports, both before and after the time of this
restriction, and even during the reign of Nero,
among other articles are named purple cloth, "fine
and ordinary," and mention is made of a number
of places where a trade in this colour was carried
on, showing such a trade to have been pretty
general.   And about the same period we find mention made of a woman Lydia being a seller of
purple, which could hardly be if the use of the
colour or tint was so restricted.   We think there
must have been some peculiarity about this restricted purple which is not well understood.

Respecting the cost and durability of the Tyrian
purple, it is related that Alexander the Great
found in the treasury of the Persian monarch
5000 quintals of Hermione purple of great beauty,
and 180 years old, and that it was worth £25 of
our money per pound weight.   The price of dyeing a pound of wool in the time of Augustus is
given by Pliny, and this price is equal to about
£32 of our money.   It is probable that his
remarks refer to some particular tint or quality
of colour easily distinguished, although not at all
clearly defined by Pliny.   He mentions a sort of
purple, or hyacinth, which was worth, in the time
of Julius Cæsar, 100 denarii (about £3 of our
money) per pound.   Again, in describing the
dye, he says that 100 lbs. of the liquor of the

T

pelagium or purpura could be purchased at about
50 denarii (30s.), the liquor of the buccinum
being double that price; also that 50 lbs. of wool
required 200 lbs. of the liquor of the buccinum,
and 110 lbs. of the purpura, to dye a durable
colour like the amethyst.   Now this would only
be about 3s. per lb. of wool; and this, at the
time when it was death to wear the colour, is not
easily reconciled.   Indeed, so much confusion
exists in the statements concerning this Tyrian
purple, that not a few have considered the whole
matter of the shell-fish dye a sort of myth; not
that there was no truth in the shell-fish producing
a dye—that cannot be gainsayed—but that the
many wonderful stories told about it in ancient
times were used as a blind to cover and conceal
the knowledge of cochineal and a tin mordant
which, it is maintained, the Tyrians possessed.
Bruce, in his " Travels," maintains this opinion,
and says, that " if the whole city of Tyre had
applied themselves to nothing but fishing, they
could not have dyed twenty yards of cloth in a
year."   And certainly, when we consider the
mode of fishing which Pliny mentions, by mussels
in a net, and the small intensity of the colour
which required 3 lbs. of liquor to 1 lb. of wool,
we would say they could not have a large trade.
Since, according to our modern researches into this
dye, one fish, the common *Purpura lapillus*, produces only about one drop of the liquor, then it

would take about 10,000 fish to dye 1 lb. of wool; so that only 3s. for this is out of the question, and even £32 is not extravagant. We think the error lies in confounding all purple colour with the Tyrian or shell-fish dye, which seems to have been rare and costly at all times, and necessarily so.

In our opinion, different methods of dyeing a purple were practised, but were kept secret. Pliny at least does not seem to have known them, and his descriptions are not very clear. Speaking of the shell-fish dye, he says the liquor of the buccinum alone gave a false dye, and that it was necessary to fix it by the liquor purpura, in order to render it durable.

The Tyrians gave the first ground of their purple dye with the unprepared liquor of the purpura, and then improved or heightened it by the buccinum. In this manner they prepared their double-dyed purple, called *Purpura debapha*, which was so called either because it was immersed in two liquors, or because it was first dyed in the wool and then in the yarn.

BLACK.—This has always been the symbol of affliction, disaster, and privation, and the colour is studiously avoided by the Orientals. Among the Jews sackcloth was an emblem of distress, and this was generally made of black hair. St John speaks of the sun becoming " black as sack-cloth of hair." Black is never referred to in

Scripture as a dyed colour. A black dress was
sometimes imposed by dominant parties upon
those who were in their power, and whom they
desired to humiliate. The Jews in Turkey are
said to be obliged to wear a black turban. In
the Greek and Roman theatres the persons whose
character the performer was enacting was generally indicated by dress. Those suffering under
misfortunes wore black, brown, or dirty white
garments. Although black in most countries has
been expressive of grief and sorrow, the Chinese
wear white as mourning; and in our own day and
nation white and black worn together are expressive of bereavement.

GREEN is not referred to in Scripture as a dye,
and very seldom in other writings. It does not
seem to have been much worn in ancient times,
probably on account of the difficulty of dyeing a
good green, which generally requires two distinct
dyes, a blue on yellow, or *vice versâ.* It is the
colour ascribed to Venus by the Sabeans, whose
astrologers ascribed seven colours to the seven
planets, Saturn black, Jupiter orange, Mars red,
Sun yellow, Venus green, Mercury blue, Moon
white. Although green and some other colours
may not have been much used as dyes, they
were common as pigments, paints, and glazes for
enamelling, and were often employed with the
same symbolical meanings. The Birs Nimroud,
or Temple of the Seven Spheres, consisted of

seven terraces, each terrace being of a distinct
colour, and these colours were symbolical of the
gods or planetary deities. The basement terrace
was black, the second orange, the third red, the
fourth golden, the fifth yellow, the sixth blue,
and the seventh silver or white, and on this last
stood the shrine, or chapel. As these different
colours were glazed or enamelled, great skill must
have been required for the work.

SPINNING AND WEAVING.

SPINNING and weaving in ancient times were principally performed by women; indeed, the words *woof*, *weaving*, and *web* are allied to the word *wife*. However, in ancient Egypt and in India men also wrought at the loom. Probably nothing could be simpler or ruder than the looms used by ancient weavers. Were we to compare these with the looms and other weaving apparatus of the present day, and reason therefrom that as the loom so must have been the cloth produced thereon, we would make a very great mistake. There are few arts which illustrate with equal force our argument in favour of the perfection of ancient art so well as this of weaving. It would appear that our advancement is not so much in the direction of quality as in that of quantity. There are few things we can do which were not done by the ancients equally perfect. Rude as were their looms in ancient Egypt, they produced the far-famed fine linen so often mentioned in Scripture and the writings of other nations. In order to show that this is not to be regarded as a merely comparative term applicable to a former age, we will here quote from G. Wilkinson respecting some mummy-cloths examined by the late Mr Thomson, of Clithero :—" My first impression on seeing these cloths was, that the first kinds were muslins, and of Indian manufacture; but this suspicion of their being cotton was soon removed by the microscope. Some were thin and transparent, and of delicate texture, and the finest had 140 threads in the inch in the warp." Some cloth Mr Wilkinson found in Thebes had 152 threads to the inch in the warp; but this is coarse when compared with a piece of linen cloth found in Memphis, which had 540 threads to the inch of the warp. How fine must these threads have been! In quoting this extract from Wilkinson to an old weaver, he flatly said it was impossible, as no reed could be made so fine. However, there would be more threads than one in the split, and by adopting this we can make cloth in our day having between 400 and 500 in the inch. However, the ancient cloths are much finer in the warp than woof, probably from want of appliance for driving the threads of the weft close enough, as they do not appear to have *lays* as we have for this purpose. Pliny refers to the remains of a linen corselet, presented by Amasis, king of Egypt, to the Rhodians, each thread of which was composed of 365 fibres. " Herodotus mentions this corselet, and another presented by Amasis to the Lacedæmonians, which had been carried off by the

Samians. It was of linen, ornamented with numerous figures of animals worked in gold and cotton. Each thread of the corselet was worthy of admiration, for though very fine, every one was composed of 360 other threads all distinct." No doubt this kind of thread was symbolical. It was probably something of this sort that Moses refers to when he mentions the material of which the corselet or girdle of the high priest was made—the fine twined linen. Jewish women are represented in the Old Testament as being expert in the art of spinning.

Ancient Babylon was also celebrated for her cloth manufacture and embroidery work, and to be the possessor of one of these costly garments was no ordinary ambition. It is not to be wondered at that when Achan saw amongst the spoils of Jericho a goodly Babylonish garment, he "coveted it and took it." The figure represented on the ancient seal of Urukh had, says Rawlinson, fringed garments delicately striped, indicating an advanced condition of this kind of manufacture five or six centuries before Joshua. It may be mentioned, however, that such manufactures were in ancient times, especially in Egypt, national. Time was of little importance, labour was plentiful, and no craftsman was allowed to scheme, or plan, or introduce any change, but was expected to aim at the perfection of the operations he was engaged in, and this led to perfection in every branch. Every trade had its own quarters in the city or nation, and the locality was named after the trade, such as goldsmiths' quarters, weavers' quarters, &c. This same rule seems to have been practised by the Hebrews after their settlement in Palestine, for we find such names in Scripture as the Valley of Craftsmen. We also find that certain trades continued in families: passages such as the following are frequent—" The father of those who were craftsmen," and " The father of Mereshah, a city, and of the house of those who wrought fine linen; " and again, " The men of Chozeba, and Joash, and Saraph, who had the dominion of Moab and Jashubalahem, these were potters, and those that dwelt among plants and hedges, and did the king's work." In ancient Egypt every son was obliged to follow the same trade as his father. Thus caste was formed. Whether this same was carried out in Babylon, Persia, and Greece, we do not know; but certainly, in these nations there were in all cases officers directing the operations, and overseers, to whom these again were responsible, so that every manufacturing art was carried on under strict surveillance, and to the highest state of perfection. As the possession of artistic work was an object of ambition amongst the wealthy or favoured portion of the community, it led to emulation among the workers. Professor Rawlinson, in his " Five Ancient

Monarchies," speaks of the Persians emulating
with each other in the show they could make of
thèir riches and variety of artistic products.   This
emulation led both to private and public exhibitions.   One of those exhibitions, which lasted
over a period of six months, is referred to in the
Old Testament; so that when we opened our Great
Exhibition in 1851, we were only resuscitating a
system common in ancient times, the event
recorded in the Book of Esther having happened
at least 2200 years before :—

"In those days, when the King Ahasuerus sat
on the throne of his kingdom, which was in
Shushan the palace, in the third year of his reign,
he made a feast unto all his princes and his
servants; the power of Persia and Media, the
nobles and princes of the provinces, being before
him : when he showed the riches of his glorious
kingdom, and the honour of his excellent majesty,
many days, even an hundred and fourscore days,
And when these days were expired, the king made
a feast unto all the people that were present in
Shushan the palace, both unto great and small,
seven days, in the court of the garden of the
king's palace; where were white, green, and blue
hangings, fastened with cords of fine linen and
purple to silver rings and pillars of marble : the
beds were of gold and silver, upon a pavement of
red, and blue, and white, and black marble.   And

they gave them drink in vessels of gold (the vessels being diverse one from another), and royal wine in abundance, according to the state of the king."

This must have been a magnificent exhibition. The number attending this feast is not ascertainable; but if the princes and nobles of the provinces (the provinces were 127 in number), and all the officers and great men of Persia and Media, and the servants of the palace, great and small, were there, it must have formed an immense company. Now, as every one drank out of a golden cup of a different pattern, we obtain an idea of profusion in art of which we can form but a very limited conception. This fact indicates that variety of pattern was an object sought after—a fashion favouring and fostering the development of art and design, and worthy of being emulated in the present day.

Speaking of the Persians, Professor Rawlinson says that the richer classes seem to have followed the court in their practices. In their costume they wore long purple or flowered robes, with loose-hanging sleeves, flowered tunics reaching to the knee, also sleeved, embroidered trousers, tiaras, and shoes of a more elegant shape than the ordinary Persian. Under their trousers they wore drawers, and under their tunics shirts, and under their shoes stockings or socks. In their houses their couches were spread with gorgeous coverlets, and their floors with rich carpets—habits that must have necessitated an immense labour and skill, and indicate great knowledge in the manu-facture of textile fabrics.

POTTERY.

No branch of manufacture presents so ancient
and close an alliance betwixt art and utility as
does the art of the potter. The date when this
art first began to show itself is lost in the dark-
ness of remote antiquity. The remains, however,
of this art which have been collected from time to
time clearly indicate a gradual advance towards
civilisation. At the same time there is no art in
which the mechanical appliances connected there-
with have begun in such perfection. The potter's
wheel is the same in construction now as it was
in the remotest antiquity of which we have any
record, the same instrument as is depicted on the
most ancient monuments. Probably it is the
most ancient mechanical appliance which industrial art has invented. Its simplicity and perfection for imparting, under expert direction, beauty
and utility to shapeless masses, is unsurpassed;
and it has descended from age to age to the
present day without any important modification.
Indeed, so absolutely identical are the modes of
working in clay, both in the mechanical contri-
vances and the manipulatory processes now in
use, and those prevalent two thousand years before

the Christian era, that were it possible to resusci-
tate the mummy of an ancient Egyptian potter,
he might immediately find employment as a first
hand in our manufacturing establishments. The
potter's wheel is a round board attached to a
lathe, and capable of being moved thereby rapidly
or slowly, as occasion may require. The round
board moves in a horizontal position, and the clay
which is to be fashioned is fixed on the centre of
it. The wheel is then put in motion, either with
the hand, or with the foot working a treadle, or by
another circular board fixed upon the bottom of
the shaft or spindle of the wheel, which is moved
by the foot of the workman himself, or, when
large and heavy vessels are being made, by a
strap and a wheel driven by an attendant upon
the potter. Mr Birch, in his "History of Ancient
Pottery," speaking about the wheel, says —
"Among the Egyptians and Greeks it was a low
circular wheel turned with the foot. Some wheels
used in the ancient Asetine potteries have been
discovered, consisting of a disc of terra-cotta,
strengthened with spokes and a tire of lead.
They are represented on a hydria with black
figures in the Munich collection, and also on a
cup with black figures in the British Museum.
The potter is seen seated on a low stool, apparently turning the wheel with his foot. Representations of the same kind are also found on
gems.

" In making vases, the wheel was used in the following manner : — A piece of paste of the required size was placed upon it, vertically, in the centre, and while it revolved, was formed with the finger and thumb. This process sufficed for the smaller pieces, such as cups, saucers, and jugs. The large amphoræ and hydriæ required the introduction of the arm. The feet, handles, necks, and mouths were separately turned or moulded, and fixed on while the clay was moist. They are turned with great beauty and precision, especially the feet, which are finished in the most admirable manner, to effect which the jars must have been inverted. The juncture of the handles is so excellent, that it is easier to break than to detach them. Great technical skill was displayed in turning certain circular vases of the class of *askoi*. With their simple wheel the Greeks effected wonders, producing shapes still unrivalled in beauty."

The modern potter generally casts a thick layer of plaster-of-paris upon his table, upon which is put the clay. As the table revolves, the workman models the clay with his finger, or by an instrument which he holds in his hand, into any kind of circular shape he may desire. We give the following quotation from Dr Thomson's " The Land and the Book," in order to show identity between the ancient and modern manner of working in the East; and we may mention that we have seen the same method of manufacture followed in potteries in this country:—" I have been out on the shore again examining a native manufactory of pottery, and was delighted to find the whole Biblical apparatus complete, and in full operation. There was the potter sitting at his frame, and turning the wheel with his foot. He had a heap of the prepared clay near him, and a pan of water by his side. Taking a lump in his hand, he placed it on the top of the wheel (which revolves horizontally), and smoothed it into a low cone, like the upper end of a sugar-loaf. Then thrusting his thumb into the top of it, he opened a hole down through the centre; and this he constantly widened by pressing the edges of the revolving cone between his hands. As it enlarged and became thinner, he gave it whatever shape he pleased with the utmost ease and expedition." Having watched this potter at work for some time, Dr Thomson adds—" From some defect in the clay, or because he had taken too little, the potter suddenly changed his mind, crushed his growing jar instantly to a shapeless mass of mud, and beginning anew, fashioned it into a totally different vessel."

The following is Jeremiah's description of the same operations :—" Then I went down to the potter's house, and behold he wrought a work on the wheel. And the vessel that he made of clay was marred in the hand of the potter, so he made

it another vessel as seemed good to the potter to
make it " (Jer. xviii. 3, 4). The art of pottery and
the products of that art are referred to in many
places of the ancient Scriptures, and spoken of also
after a manner which makes it evident that pottery
was a common art, and extensively practised by
the Hebrew people. There are descriptions of the
preparation of the clay : it was dug, then trodden
by men's feet, after which it was given to the
potter. From a passage in 1st Chronicles it seems
evident that there were among the Israelites something like national manufactories of pottery ; at all
events there were manufacturers to the king, and
the art was pursued in these works by the descendants of a certain family. "The sons of Shelah
the son of Judah were Er, the father of Lecah, and
Laadah, the father of Mareshah, and the families
of the house of them that wrought fine linen, of
the house of Ashbea, and Jokim, and the men of
Chozeba, and Joash, and Saraph, who had the
dominion in Moab, and Jashubi-lehem. And these
are ancient things, these were the potters, and
those that dwelt among plants and hedges : there
they dwelt with the king for his work " (1 Chron.
iv. 21–23).

The ease with which clay could be formed into
any shape offered a ready means for the development of art; and the ease with which it took
impressions and retained these when dried, rendered
clay generally useful for many purposes. Earthen

U

vessels were used both by the Egyptians and other
nations for culinary purposes; tiles with patterns
and with writing upon them were common, and
were used as money and for deeds, which were
sometimes put into earthen vessels for preservation,
and the vessel then buried.   A practice common
with many nations of antiquity was to put the
ashes of the dead into earthen vessels, and bury
them; in some cases the entire body was put into
jars.   Enormously large vessels were made of clay,
by fitting up a wooden frame-work and laying
clay over it.   In burning or baking the clay the
wooden frame was destroyed, and a vessel left of a
size sufficient to form a residence for a person.   It
was in one such that Diogenes lived when visited by
Alexander the Great.   Such vessels also served as
casks for holding large quantities of wine, and they
were used also for honey and figs.   Great skill
and attention was required in making them; this
attention was probably more requisite in the dry-
ing and baking than in building.   This at least is
our experience at the present day when making
large clay vessels, such as those used for the
manufacture of glass.

The discovery of the method of glazing, painting, and enamelling, must have been a great step
onward in the art of pottery; and such glazed
vessels are found among the ruins, and in the
tombs of people who lived in far remote antiquity.
If, as has been often remarked, we can estimate

the state of refinement of a people by the kind
and quality of their pottery, then most of the
ancient nations must have reached a high state
of refinement. As it is with respect to some
other practical arts, our highest effort now is
to equal ancient works, so is it in regard to the
potter's art—our highest effort is to equal ancient
works. To surpass such works seems a position
too high to attempt, although within the present
century much has been done towards a high
position in the fictile art. Mr Birch is of
opinion that the Jews did not do much in the art
of pottery : he says, "No remains of earthen
vessels used by the Hebrews, or even of bricks
employed in the construction of edifices, are
known; the pottery which is occasionally found on
the site of Jerusalem being principally either the
red Roman ware, or that called Samian. The
depth of *debris* which in some places reaches 40
feet, and the fact of no excavations having been
undertaken on the site, are the probable reasons
why entire vases or other Terra-cotta objects have
not been discovered, whilst the low state of the art
among the Jews may have caused the fragments,
which must always abound in the vicinity of great
cities, to be neglected. It is, however, possible
that the Jews obtained the principal earthenware
they required from Egypt, and that, as among
other Oriental peoples, metallic vessels were pre-
ferred for the kitchen or the table." There is, no

doubt, truth in this last remark, and we have else-
where shown that the Jews did use metallic
vessels in their houses, but that earthen vessels
were also in common use is evident from several
passages of Scripture, and what we have quoted
from 1st Chronicles is positive evidence that there
were also manufactures of pottery in Judea in the
time of Solomon.   We may here direct attention to several passages in Scripture referring to
pottery, which are suggestive and interesting
as regards social habits.   The potter's vessel is
made the symbol of weakness and fragility, and
when the prophet speaks of the wicked being
broken as a potter's vessel, he symbolises their
utter destruction.   Jars and such vessels were
made extremely thin, and were consequently more
fragile than are most of our common stone or
delf ware.   Isaiah says (xxx. 14), "He shall
break it as the breaking of a potter's vessel that
is broken in pieces: so that there shall not be
found in the bursting of it a shred to take fire
from the hearth, or take water withal out of the
pit."   In connection with this Dr Thomson says :
"It is very common to find at the spring or pit
pieces of broken jars, to be used as ladles either to
drink from or fill with, and bits of fractured jars
are preserved for this purpose.   Take your stand
near any of the public ovens in Sidon in the
evening, and you will see the children of the poor
coming with 'shreds' of pottery in their hands,

into which the baker pours a small quantity of hot embers and a few coals with which to warm up their evening meal. Isaiah's vessels, however, were to be broken into such small bits that there would not be a shred of sufficient size to carry away a few embers from the hearth." These comparisons are exceedingly interesting. It has been supposed by some commentators on Scripture that owing to the abundance of fragments which lay strewn around the royal establishment of potters near Jerusalem, this locality was named the potters' field, and that it was this field that was bought with the thirty pieces of silver given to Judas for betraying his Master. If this suggestion can be proved, it would show that pottery was a manufacture carried on by the Jews from the time of their greatest prosperity under Solomon, until the time when Judea was made a Roman province, and the national or royal pottery abandoned; and this waste field, that had been in connection with it, was bought from their rulers and used as a burial place for strangers.

GLASS.

GLASS, according to Pliny (Hist. Nat. xxxvi. 26), was discovered by what is termed accident. Some merchants kindled a fire on that part of the coast of Phœnicia which lies near Ptolemais, between the foot of Carmel and Tyre, at a spot where the river Belus casts the fine sand which it brings down; but as they were without the usual means of suspending their cooking vessels, they employed for that purpose bags of nitre—their vessel being laden with that substance: the fire fusing the nitre and the sand, produced glass. The Sidonians, in whose vicinity the discovery was made, took it up, and having, in process of time, carried the art to a high degree of excellence, gained thereby both wealth and fame. Other nations became their pupils; the Romans, especially, attained very great skill in the art of fusing, blowing, and colouring glass.

Whether the story of these merchants be strictly correct we cannot say, but certainly if nitre came into contact with a fire of glowing charcoal on a bed of sand, there were all the conditions for forming glass. By intense heat the nitre would be converted into the carbonate of potash, which unites with great facility with sand. The date of this discovery is nowhere mentioned, but it must have been very far back; for glass was in use before the time of Moses, and in all probability Moses refers to this manufacture, or at all events to the sand used for it, in his blessing of the tribe of Zebulun, whose borders were unto Sidon. He says, "They shall suck of the abundance of the sea, and of treasures hid in the sand." Notwithstanding this statement of Pliny, which is also corroborated by Strabo and other more ancient writers, it was long denied that the ancients were acquainted with glass, properly so called; nor did the denial entirely disappear even when the city of Pompeii proved it to be without foundation.

Sir E. B. Lytton says, "The discoveries at Pompeii have controverted the long-established error of the antiquaries, that glass windows were unknown to the Romans. The use of them was not, however, common among the middle and inferior classes in their private dwellings." In the tepidarium of the public baths, a bronze lattice came to light with some of the panes still inserted into the frame.

Wilkinson, in his "Ancient Egyptians," has adduced the fullest evidence in proof that glass was known to, and made by, that ingenious people at a very early period of their national existence. Upwards of 3500 years ago, in the reign of the first Osirtasen, they appear to have practised the art of blowing glass. The process is represented in the paintings of Beni Hassan, executed in the reign of that monarch. In the same age images of glazed pottery were common. Ornaments of glass were made by them about 1500 years B.C., for a bead of that date has been found, of the same specific gravity as that of our own crown glass. Many glass bottles, &c., have been met with in the tombs, some of very remote antiquity. Glass vases were used for holding wine as early as the Exodus. Such was the skill of the Egyptians in the manufacturing of this article, that they successfully counterfeited the amethyst and other precious stones. Winckelmann is of opinion, that glass was employed more frequently in ancient times than in modern. It was sometimes used by the Egyptians even for coffins. They also employed it, not only for drinking utensils and ornaments of the person, but for Mosaic work— the figures of deities and sacred emblems being of exquisite workmanship, and a surprising brilliancy of colour. The art, too, of cutting glass was known to them at the most remote period.

Glass articles, made for ornamentation and for works of art, which, at the present day, we cannot rival, let alone surpass, have been found amongst the ruins of Egypt; and so common was its use that necklaces, head-dresses, and other common ornaments, enhanced with glass bugles and beads, were worn by the common people, and imitation jewellery, made of glass of different colours, was worn then as now, and so perfect were these imitations of the precious stones that it required the most expert specialists to detect the difference. Indeed, glassmaking in Egypt must have been a very extensive manufacture; therefore, had we no other proof, we are justified in supposing that the Hebrews, while in Egypt, obtained a knowledge of this art, and brought it with them. This being the case, the reference to the value of sand, in the blessing of Zebulun, would be fully appreciated. Job speaking of wisdom says, "The gold and the crystal cannot equal it." Many scholars say that the word translated crystal should be glass; and there were kinds of glass that were more precious than gold.

Pliny describes the district which fell to the lot of Zebulun as being the seat of glassmaking. Moses probably refers to the same fact when he speaks of treasures in the sand. This supposes on his part a knowledge of glassmaking, and we have seen that the Egyptians were expert glass manufacturers long before the time of Moses. Figures, found upon the tombs in Egypt, exhibit men working at the art, using blowpipes exactly as they are used at the present day. They possessed the art of introducing various colours homogeneously into the same vessel, a thing which glassmakers at the present day are still unable to do. Mr Wilkinson states that, a few years ago, the glassmakers of Venice made several attempts to imitate the variety of colours found on antique cups, and failed. Another curious process common in ancient times has also, and only, been attempted at Venice. In this case, the pattern on the surface passes in directly through the substance, so that if any number of sections be made, each section has the same device on its outer and inner, or upper and under surface, such work being, in fact, a glass mosaic. The skill displayed in this work lies not merely in the process, but in the fineness of the design, for some feathers of birds, and other details in the figures, are only to be made out with the aid of a lens. This fact indicates that means of magnifying were also known, and employed, by workers in ancient times, for certainly we would require to use such aids in order to produce the same work. There is one piece of glass mosaic referred to by Wilkinson, representing a bird, concerning which it is. said, that " the most delicate pencil of the miniature artist could not have traced with greater sharpness the circle of the eyeball and the plumage of the neck and wings." It is humbling to our vanity to find that we, with our boasted skill in arts, are yet behind our fellow-workmen who lived 4000 years ago. The purpose to which such wonderful workmanship was applied helps us to

apprehend the social condition of the people among
whom it flourished. It takes a long time for a
people to acquire such an exquisite artistic taste,
that nothing but the very finest class of workmanship will be worn as ornament. A rude, ignorant
people hang fragments of broken glass around
their necks, and are satisfied with its mere bright-
ness or transparency;. not so a cultivated people.
The wealthy pay for and procure the highest class
of workmanship and material, the poor emulate
them, and ingenuity is thus set a-thinking to
devise and produce imitations in cheaper material
to supply the desires of the people. Wilkinson
found in Thebes counterfeit pearls which, even
with a strong lens, it was difficult to detect to
be impositions. Then, probably better than now,
a working man's wife or daughter might purchase
from an Egyptian dealer in ornamental goods a
necklace of imitation stone, not much inferior in
look to the real. Necklaces made with glass
bugles and beads, such as are common at the
present day, and nets made with different coloured
beads, of similar workmanship with our bead
purses, were worn by the Egyptian beauties when
Abraham and Sarah went down into that country.

Ancient profane authors make mention of immense emeralds, which are considered now to have
been made of glass. An emerald was presented
by the King of Babylon to an Egyptian Pharaoh,
which was six feet long and four and a half broad;

and an obelisk composed of four emeralds, in the Temple of Jupiter, was sixty feet high and six feet broad. To form pieces of glass of such a size, having the appearance of an emerald stone, was a triumph of skill which certainly cannot be surpassed at the present day, and, if produced, would yet be a fit present to send to a king.

The art of engraving on glass was also well known in ancient times; but we have referred to this art in connection with the engraving of the names of the tribes of Israel on the stones in the breastplate. It is said the engraving was like the engraving of a signet. Layard states that travellers can scarcely walk any distance in the neighbourhood of the country where Nineveh and Babylon are supposed to have stood without striking his feet against a glass fragment. Among the excavations made in the mounds, Dr Layard discovered some entire glass bowls. One of these has on it cuneiform characters, which enables us to fix its date at the latter end of the seventh century B.C. And he says, "It is the most ancient piece of transparent glass the date of which has been ascertained." We do not see how imitations of precious stones could be made in glass without transparency, however the dates of these imitations have not been ascertained. Professor Rawlinson says, in his "Five Ancient Monarchies," that transparent glass was first brought into use, or at least the oldest specimen found is, in the reign of Sargon, 710 B.C. ; but says that lenses of transparent glass must have been in use before this on account of the minuteness of the work in their engravings, which supposition is confirmed by the discovery of a lens at Nimrod.

The ancient Assyrians cut gems with great skill. Layard mentions the discovery of a lens, but says it was of rock crystal, so that it is possible they may have had lenses for engraving before transparent glass was used for that purpose. Still the discovery of a lens, whatever was its substance, is most important, for it shows that the ancients were acquainted with the laws of light, and the concentration of its rays. They undoubtedly knew practically much which they may not have understood scientifically. There was found in Pompeii a convex glass for a lamp, which remained in the wall, dividing two apartments in the public baths near the forum. From analyses of ancient coloured glasses and enamels, the colouring matters are found to have been for—

Bright Red	. Oxide of Lead (Red Lead)
Dull Red .	. Oxide of Iron
Purple Red	. do.
Rich deep Red .	Suboxide of Copper
Yellow .	. Yellow Ochre and Chalk
Orange .	. Yellow Sulphide of Arsenic
Blue. .	. Oxide of Copper, Alumina, Lime ;

depth of tint depending on the quantities of these used.

BUILDING.

THE building art meets us on the very threshold
of history. We read that Cain built a city and
called it after the name of his son. Shortly after
the Flood also, when man may be supposed to have
begun to work up the arts of civilisation anew, we
read of the project of building a tower, the top of
which would reach to heaven. The very conception
of such a work as this indicates a feeling of great
power and ability. Such an idea could not proceed
from a low uncultivated mind, nor from people who
had no experience in art. The materials they had
to work with, and the method of preparing these
materials, all indicate skill. "And they said one
to another, Go to, come let us make brick, and burn
them thoroughly, and they had brick for stone and
slime had they for mortar." In connection with
this tower, they also intended to build a city.
"They said, Go to, now, and let us build us a city
and a tower." How far these projected buildings
were proceeded with cannot now be determined. The
general opinion is, that they were stopped very
early; however, that is only a surmise. Many
exaggerated stories have been told concerning this
great tower, owing no doubt to misunderstand-
ing. Travellers finding an extensive tower of

great height near Babylon concluded that it was the
remains of the Tower of Babel, pictures of which,
showing a high tower with a spiral road leading
upwards round it, may be seen in old illustrated
Bibles, which was calculated to have been about
one fourth of a mile in height. However, more
recent and accurate research has shown that the
ruin now standing was built by Nebuchadnezzar,
many centuries after the date of the projection of the
Tower of Babel. On this matter, and on the nature
of the buildings in Babylon and Chaldea, we will
quote from Prof. Rawlinson's "Five Ancient
Monarchies."

"In mineral products Chaldea was very deficient
indeed. The alluvium is wholly destitute of metals
and even of stone, which must be obtained, if
wanted, from the adjacent countries. The neighbouring parts of Arabia could furnish sandstone,
and the more distant parts basalt, which appears to
have been transported occasionally to the Chaldean
cities. The Chaldeans found, in default of stone,
a very tolerable material in their country which
produced an inexhaustible supply of excellent clay,
easily moulded into brick, and not even requiring to
be baked in order to fit it for building. Exposure
to the heat of the summer sun hardened the clay
for most purposes, while a few hours in a kiln
made it as firm and durable as freestone or even
granite. Chaldea again yielded various substances
suitable for mortar. Calcareous earths abound on

the western side of the Euphrates towards the
Arabian frontier, while everywhere a tenacious
slime or mud is easily procurable, which, though
imperfect as a cement, can serve the purpose, and
has the advantage of being always at hand.
Bitumen is also produced largely in some parts,
particularly at Hit, where are the inexhaustible
springs which have made that place famous in all
ages. Naphtha and bitumen are here given forth
separately in equal abundance, and these two
substances, boiled together in certain proportions,
form a third kind of cement superior to the slime
or mud, but inferior to lime mortar."

In describing the remains of the *Mugheir* Temple
he says, " The material of which its inner structure
is composed seems to be wholly or partially burned
brick of a light red colour laid in a cement composed of lime and ashes. This central mass is faced
with kiln-dried bricks of large size and excellent
quality, also laid, except on the north-west face, in
lime mortar. The north face is laid with bitumen.

" The earliest traditions and existing remains
of the earliest building alike informs us that the
material used was brick. The earliest brick hither-
to discovered in Chaldea is $11\frac{1}{4}$ inches square by
$2\frac{1}{4}$ inches thick; the baked brick of later date is
longer, being 13 inches square by 3 inches thick.
The sun-dried brick was laid in mud as mortar;
for the kiln-dried, bitumen was used. The next
king in succession is Nimrod, the mighty hunter,

and the earliest monarch of whom any remains have been obtained. Not only are his bricks found in a lower position than any others, at the very foundation of building, but they are of a rude and coarse make; and the inscriptions upon them contrast most remarkably in the simplicity, in the style of writing used, and in their general. archaic type, with the elaborate and often complicated symbols of the later monarchs. The style of *Urukh's* building is also primitive and simple in the extreme. His bricks are of many sizes and ill fitted together; he belongs to a time when even the baking of bricks seems to have been comparatively rare, and he is altogether unacquainted with the use of lime mortar, for which his substitute is moist mud or bitumen; his age is as early as 2326 B.C., *possibly a century earlier.*" The author further says, "The works of this ancient monarch are very gigantic and numerous; the base of one mound at Warka is 200 feet square, and about 100 feet high. The buildings of this king seem to have been all designed for temples. They are carefully placed with their angles facing the cardinal points, and are dedicated to the sun, the moon, to Beltis (Bel-Nimrod). However, rude as were the materials of his building, there is evidence of great advancement in art. Speaking of the monarch's signet, which consisted of a metal frame on which was mounted a precious stone, Professor Raw-

x

linson says, " So far as we can judge from the
representation, which is all that we possess of this
relic, the drawing on the cylinder was as good,
and the engraving as well executed, as any work
of the kind either of the Assyrian or of the later
Babylonian period."

In reference to the Tower of Babel, Professor
Rawlinson, in Smith's "Dictionary of the Bible,"
says, " When Christian travellers first began to
visit the Mesopotamian ruins, they generally
attached the name of the ' Tower of Babel ' to
whatever mass, among those beheld by them, was
the loftiest and most imposing.   There are in
reality no real grounds, either for identifying the
tower with the temple of Belus, or for supposing
that any remains of it long survived the check
which the builders received when they were
scattered abroad upon the face of the earth, and
' left off to build the city.'   All, then, that can
be properly attempted by the modern critic is to
show, *first*, what was the probable type and character of the building, and *second*, what were the
materials and manner of its construction."   The
author considers that although the ruin called
Birs-Nimrud cannot be the Tower of Babel
referred to in Scripture, it represents the form of
Babylonian temple towers, and describes it thus :—
" Upon a platform of crude brick, raised a few
feet above the level of the alluvial plain, was
built of burned brick the first or basement stage,

an exact square 272 feet each way, and 26 feet in perpendicular · height. Upon this stage was erected a second 230 feet each way, and likewise 24 feet ¦high, which, however, was not placed exactly in the middle of the first, but nearer to the south-western end, which constituted the back of the building. The other stages were arranged similarly; the third being 188 feet square, and 26 feet high; the fourth, 146 feet square, and 15 feet high; the fifth, 104 feet square, and 15 feet high; the sixth, 62 feet square, and 15 feet high; the seventh, 20 feet square, and the same height. On the seventh stage there was probably placed the ark or tabernacle, which seems to have been again 15 feet high, and must have entirely or nearly covered the whole seventh story. The entire original height, allowing three feet for the · platform, would thus be 156 feet. The whole formed a sort of oblique pyramid."

Loftus, in his " Ancient Chaldea," in reference to the *Birs*, states, that the different terraces are of different colours, and that these may have been produced by adding to the clay some chemical compound before the bricks were burned; and then adds, " It is more difficult to explain the cause of the vitrification of the upper building. Captain Newbold originated an idea when we examined the Birs-Nimrud in company, which is now beginning to be adopted, that in order to render their edifices more durable, the Babylonians submitted

them when erected to the heat of a furnace. This
will account for the remarkable condition of the
brickwork on the summit of the Birs-Nimrud, which
has undoubtedly been subjected to the agency of
fire."

If the surface of the building was vitrified after
the building was erected, it is certainly somewhat extraordinary when we consider the size and
the height. It would require very great care and
skill to effect vitrification over so large a surface, and
at the same time prevent fusion and running of
the fused mass at the hotter parts of the fire.
We have had samples of the bricks brought from
the Birs by Mr Fulton, who visited it a few
years ago, sun-dried, kiln-baked, and vitrified,
and having subjected them to analyses, found that
they were all composed of the same kind of clay.
The following are the analyses of the kiln-baked
and vitrified samples—

	KILN-BAKED.	VITRIFIED.
Silica,	48·30	50·88
Peroxide iron,	23·93	24·46
Alumina,	5·50	5·87
Lime,	20·07	17·46
Magnesia,	·94	1·04
Moisture and loss,	1·26	·29
	100·00	100·00

The colour of the kiln-baked was a light yellow,
resembling in appearance bath-brick, but hard.
This composition would be very easily vitrified,
requiring only a moderate heat and time. If this

heat was raised to high intensity, such a compound would easily fuse and run down the face of the building : even in burning the bricks, a high heat would soon render them shapeless. As the bricks still retain the king's mark sharply stamped upon them, great care must have been taken in regulating the heat. One circumstance which militates against Captain Newbold's theory of vitrification after the building was erected, is the mortar, it being of bitumen. A vitrifying heat would set fire to the mortar and destroy the edifice. The Babylonians were also acquainted at a very early period with the arch, for some regularly constructed semicircular arches have been discovered among the ruined building. Pieces of baked clay also, with inscriptions imprinted upon them, were used as a circulating medium in the same way as our bank-notes, these representing certain amounts of gold and silver, and were payable at the royal treasury. These brick bills date so far back as 600 B.C. Not only, therefore, were these Babylonians skilled in art, but were far advanced in political and social economy.

In countries where stone was accessible it was used for building, and evidence exists of the great perfection attained in early ages in the art of building with stone. Ancient Egypt, still a mystery, has left ample evidence of her great skill in the building arts. The Great Pyramid, which is yet an enigma, stands for our astonishment. Herodotus tells us, when speaking of the Labyrinth of Egypt, that it had 3000 chambers, half of them above and half below ground. He says, "The upper chambers I myself passed through and saw, and what I say concerning them is from my own observation. Of the underground chambers I can only speak from the report, for the keepers of the building could not be got to show them, since they contained, as they said, the sepulchres of the kings who built the labyrinth, and also those of the sacred crocodiles; thus it is from hearsay only that I can speak of the lower chambers. The upper chambers, however, I saw with my own eyes, and found them to excel all other human productions. The passage through the houses, and the various windings of the path across the courts, excited in me infinite admiration, as I passed from the courts into the chambers, and from chambers into colonnades, and from colonnades into fresh houses, and again from these into courts unseen before. The roof was throughout of stone like the walls, and the walls were carved all over with figures. Every court was surrounded with a colonnade, which was built of white stone exquisitely fitted together. At the corner of the labyrinth stands a pyramid forty fathoms high, with large figures engraved on it, which is entered by a subterranean passage." No one who has read an account of the Great Pyramid of Egypt, the building of Solomon's Temple, and of the ruins of ancient stone buildings still remaining, will doubt the ability of the ancients in the art of building with stones.

In connection with the subject of stone buildings, and indeed buildings of all kinds, a very important question suggests itself; namely, What kinds of mortar were used? There is a clear statement concerning brick buildings, that the adhesive materials employed were slime and bitumen. The former was, no doubt, mud or thin clay similar to that used for making the bricks, so that sun-dried bricks, built with this, would form a solid clay mass. The bitumen was used with the kiln-baked brick formerly referred to, and was made somewhat similar to that we use for pavement. We analysed a sample of this asphalt which we found adhering to the sample of brick which we had from Birs-Nimrud, and found that it was composed of 25 per cent. of bitumen, and 75 per cent. of ground limestone, not finely ground, as there were pieces in it about the size of a split pea, which enabled us to determine that the lime had not been previously burned, but was common grey limestone. The composition was both hard and tough, and resembled in character *val de travers*. The question may be asked, Why should they have taken the trouble to grind limestone when coarse sand would have served the purpose? Doubtless they used what had been found in practice to be the best, and a building, made with bricks " thoroughly burned,"

and laid in bitumen composed as above, was well
fitted to stand for centuries; and if, as is supposed
by Captain Newbold, the whole surface of the
Birs-Nimrud was vitrified, such vitrification would
certainly secure it against all atmospheric influ-
ences.   However, we cannot coincide with the
opinion that the vitrification was an intentional
act, performed when the building was being built,
for the application of fire to the building would
kindle the mortar, and bring the whole face of
the building into active combustion, and destroy
the beauty of the surface, if not the stability of
the entire building. Probably the vitrification seen
upon some portions was done by the firing of the
mortar accidentally.   We have seen samples of
enamelled brick from Seleucia; they had been
baked, and were made of the same material as
the brick of Birs-Nimrud, the enamel was a beautiful dark-purple colour. It had evidently been
put on in a liquid state, the liquid being thick
and pasty, as none of it had penetrated the porous
material of the brick.   It had afterwards been
vitrified, probably in an oven.   The enamel formed
a thin glass over the surface of the brick of about
the thickness of ordinary writing paper.

We have searched with care many archæological
works, in the hope of finding some definite information concerning the kinds of mortar used by
ancient stone builders, but the statements of the
various travellers and excavators who have visited

ancient seats of civilisation are not very satisfactory.
They refer without discrimination to *mortar, lime,
lime-mortar, cement, plaster, &c.*, but in few cases
do they specify the particular nature of the mortar
or cement of which⋅ they speak. Loftus says,
" The vaulted roof of the houses and mosques at
Mosul are constructed of gypsum, plaster, and
broken brick, the terrace being covered with mud
and earth; such may have been the case in the
palaces of ancient Nineveh. The numerous fragments of brick and lumps of decomposing gypsum
in the soil above the sculptures are strong presumptive evidence that this plan of constructing their
roofs was adopted by the Assyrians." Similar
indefinite statements with regard to mortars and
cements are made by Layard and others. In
Smith's "Dictionary of the Bible," a passage in
Isaiah, in which the prophet speaks of the people
being as the burnings of lime, is adduced as
evidence that the ancient Jews were acquainted
with quick-lime as a mortar. The context, however,
gives no warrant for such an opinion. The burning
of the bones of a king into lime, is also adduced
as evidence, which it is not. Dr Thomson, in
" The Land and the Book," says, that "The houses
in Palestine are built without lime-mortar, and
that the mortar used, during rain, becomes as
slippery as soap, and if constant care is not taken
the houses soon fall." Job speaks of houses
which are not inhabited soon becoming heaps.

Solomon also refers to the rapid decay of houses through slothfulness, and Ezekiel speak of walls that had been built with untempered mortar as soon falling. No doubt these references are to ordinary buildings as dwelling-houses, but they show that mortar such as we now use was not used then for ordinary structures.

Dr Wallace of Glasgow, in a paper on ancient mortars, gave analyses of samples taken from the Great Pyramid :—

	INTERIOR.	EXTERIOR.
Hydrated sulphate of lime . .	81·50	82·89
Carbonate of lime . . .	9·47	9·80
Carbonate of magnesia . .	·59	·79
Oxide of iron	·25	·21
Alumina	2·41	3·00
Silicic acid '	5·30	4·30
	99·52	100·99

In the work of J. J. Vicat on mortars, there is an analysis of a sample from the Great Pyramid differing little from that by Dr Wallace :—

Sulphate of lime with a little alumina . .	70·30
Carbonic acid and lime	16·34
Sulphate of soda	3·20
Water and loss	10·16
	100·00

These have no relation whatever to our mortars, but consist simply of plaster of Paris with certain

impurities. The description by Pliny of the preparation of plaster of Paris accounts · for such
impurities. He says, "There are numerous varieties
of it: one kind is prepared from a calcined stone
as in Syria. In Cyprus, gypsum is dug out of the
earth, the stone that is calcined for this purpose
ought to be similar to the alabastrites or else of
a grain like that of marble. In Syria they select
the hardest stones for the purpose, and calcine
them with cow-dung to accelerate the process.
Experience has proved that the best plaster of all
is that prepared from specular stone, or other
stone similarly laminated. Gypsum when moistened must be used immediately, as it hardens
with the greatest rapidity. It is useful for ornamental figures and wreaths in buildings." Here we
find plaster applied to the same purposes to which
we now apply it, but not as mortar. The process
of calcination described by Pliny would naturally
lead to the knowledge of quick-lime, and he says,
"It is truly marvellous that limestone after being
subjected to fire should ignite on the application
of water." In all probability slaked lime was
used in the first instance as a sort of putty. Dr
Wallace gives an analysis of two samples, one
from the Pnyx, the other from an ancient temple
at Pentelicus near Athens. The mortars are very
old, but no date can be assigned. The analyses
are—

	PNYX.	TEMPLE.
Lime	45·70	49·65
Carbonic acid . .	37·00	38·33
Magnesia . . .	1·00	1.09
Sulphuric acid . .	...	1·04
Peroxide of iron . .	·92	·82
Alumina . . .	2·64	·98
Silicic acid and sand .	12·06	3·90
Water	·36	3·07
	99·68	98·88

Dr Wallace also gives the analysis of a specimen found in Cyprus, which was used as a cement for joining water-pipes placed ten feet below the surface of the ground. Its composition was—

Lime	51·58
Carbonic acid . . .	40·60
Magnesia	·70
Sulphuric acid . . .	·82
Alumina	·40
Silicic acid and fine sand .	·96
Organic matter . . .	·24
Water	3·09
	98·39

This is as near as possible the same in composition with that taken from the temple near Athens. Neither are the same as our modern lime mortar.

Mr Vicat gives three separate analyses of mortars taken from structures erected in France by the Romans during the reign of Agrippa. These are the earliest examples we have found of the use of our modern lime-mortar—

	FROM A TOWER. 1ST CENTURY.	REIGN OF AGRIPPA. 1ST CENTURY.	
Lime	. 29·1	15·16	24·00
Carbonic acid . .	. 20·0	9·00	12·00
Silicic . . .	. 45·0	68·25	56·25
Water . . .	. 5·9	4·00	5·00
Alumina and iron-oxide .	...	2·75	2·75
	100·0	99·16	100·00·

The first of these is much richer in lime than the mortar we now use, and the lime and sand must have been very pure. The two others are of much the same composition as the lime-mortar used at the present time. Marcus Portius Cato, who lived about two centuries before our era, describes the method of making lime-mortar in his day, which is the same as we now follow. He states the proportions of lime and sand employed, and describes the outward properties which characterise a good limestone, and also mentions the construction of lime-kilns. Vitruvius, who lived about 200 years later, gives more details, speaks of sea sand not being so well adapted for mortar as that obtained from pits, on account of its taking longer to harden, and that pure limestone was more serviceable. He was the first to notice the properties possessed by Puzzolanas, a volcanic product with the wonderful property of causing lime to set rapidly under water, and to be used for harbour construction, and for structures under water.

The probability is that lime putty was originally

used, mixed with small stones to form a concrete, and was used in this way before the practice of mixing sand with lime came into vogue. It is the opinion of many who have studied the subject, that in the more ancient stone buildings no mortar of any kind was used, or if any, only a thin layer of plaster of Paris. This is the cement used in the building of the Great Pyramid. In a book just published by the committee of the Palestine Exploration Fund, entitled "Our Work in Palestine," in reference to the ancient wall of the temple in Jerusalem, the following passage occurs :—" The masonry of this wall presents several marked and very important differences of work. These, as we shall shortly see, may be divided into five. The stones are thus prepared : In the first instance they are dressed square on the upper and under surface and at the two ends. The dressing is in many cases so true that a knife cannot be inserted between two stones ; they are placed one above another, each stone being set half an inch to an inch further back, so that the wall is not perpendicular, but stands at a slight angle, the great advantage being that buttresses and other supports are not needed. No mortar or cement has been used." In an archæological magazine, *Long Ago*, for February 1873, appears the following in reference to buildings supposed to be of the same date as Solomon's Temple.

M. Mauch, an African traveller, is of opinion that he has discovered the Queen of Sheba's palace, about which he thus writes: " The ruins which have been so often spoken about, are composed of two masses of edifice in a tolerably good state of preservation. The first is on a mountain of granite; and amongst other constructions is to be remarked one which is an imitation of Solomon's Temple, being fortress and sanctuary at the same time, the walls of which are built in wrought granite without mortar, and still being more than 30 feet high; beams of cedar served as ceiling to the narrow and covered galleries. No inscription exists, but only some special designs of ornamentation which announce a great antiquity. The whole western part of the mountain is covered with blocks of great size, which seem to indicate terraces. The second mass of ruins is situated to the south of the mountain, from which it is separated by a low valley; it retains a well-preserved circular form, with walls constructed as a labyrinth, also without mortar. A tower still exists, 30 feet high, 17 feet diameter at base, and 9 feet at top. The circular is accompanied by a large number of others, situated in the front, and which, doubtless, served as the habitation of the Queen of Sheba's suite. I was confirmed by the natives themselves in the idea that these ruins date from the Queen's time. The natives still call the circular building the house of the great princess."

Speaking of Solomon's Temple, the authors of "Our Work in Palestine" say, that an examination of the ancient wall presents no signs of chippings or of stonedressing having been performed in its neighbourhood, which signs would have been present had the stones been prepared at the spot. How well this agrees with the Scripture account, " And the house, when it was in building, was built of stone made ready before it was brought thither, so that there was neither hammer nor axe, nor any tool of iron, heard in the house while it was building."   Again, in connection with excavations made on Mount Gerizim, where there are two ruins, the one known as the castle, the other as the church, it is stated, " The church and castle were found to be built on a rough platform of large stones, laid together without mortar." We wrote to Captain W. Wilson, R.E., one of the exploring expedition.   In the letter which we received in reply to our inquiry, he says, " I do not remember having seen mortar used in any of the very early buildings in Palestine.   The custom seems to have been to dress the beds and joints of the stones very finely, and set them without mortar."   Piazza Smith, in his " Work at the Great Pyramid," says, " The stones fit so closely upon each other that one cannot insert the edge of a penknife between them."   This practice of so preparing each stone to fit exactly its proper place, is suggestive of a great attainment in architectural skill. In the case of such erections as the Temple of Solomon, for which the stones were prepared in the quarry at a distance from the building, the architect must have had a detailed plan of the building, showing the exact dimensions of every stone with its place in the building.

While speaking of the temple, we may relate a story which we received as part of our instructions when we joined the Freemasons. It has reference to the corner-stone mentioned in Scripture, and the substance of our instructions was as follows: —The corner or finishing stone in ancient temples was generally regarded as a sacred stone, and had certain sacred markings upon it. It was laid with great ceremony, and was supposed to confer a sacred influence on the building. David began early to make preparation for the building of the temple; and among the things which he had in preparation was a massive stone of an angular shape without any carvings or marks of any kind upon it. This stone was carried to the ground where the temple was to be erected at the commencement of the building operations, and the instructions to the builders were, that it was to be taken care of, and carried up with them, as the building proceeded, until its place was found. This the builders considered a sort of useless labour, and were often inclined to throw it over; but on coming to the finish of their work, when they were expecting the grand sacred corner-stone to be brought

Y

to them, the stone which the builders rejected became the head of the corner, the space left was found to fit this stone exactly, and Solomon, like a true master mason with mallet in hand, may have repeated the cheering words, It is finished.

There is often mention made of large pools and reservoirs, stone-built erections for holding water. We cannot conceive that the stones for these could be so fitted as to be water-tight without cement, but of what kind was the cement used we have no positive evidence. Travellers do not seem to have paid any attention to these matters. Captain Wilson says in the letter referred to previously, " Cement is used in the aqueducts and pools, and an analysis of these would be interesting, but it is difficult to say at what time the coating has been put on, the constant repairs, &c., having since obliterated the old work." However, stones prepared with the exactness here indicated, and laid with thin plaster of Paris, or lime putty between them, would form a tight wall for a reservoir, but when limestones were used for the building of these reservoirs, they might ultimately require repair. Baldwin, in his " Prehistoric Nations," says that " Mr Wellsted, speaking of ruins he saw at Nakab-el-Hadjar, says that he found the remains of an immense wall originally from thirty to forty feet high, and ten feet thick at the foundation. The blocks of greyish marble of which it was built were hewn and fitted by the builders with surprising nicety, indicating science and skill in construction of a high order. The magnitude of the stones used, and the perfect knowledge of the builders, would give the structure importance in any part of the world." And speaking of the ruins, he says, " Their origin belongs to a very remote antiquity." Concerning ruins found at El-Belid, he says, " The blocks of stone used by the architects, cut with geometrical precision, show marvellous perfection in the workmanship of the builders." Researches concerning the state of the building arts in ancient Egypt, and among the ancient Arabians, or Cushites, discover to us that the further back we go, the more perfect is the art. This art like others must have had a gradual development, and the interval between the perfection of the art and the first beginnings must reach back to a very remote time.

THE ARK OF NOAH.

In the 6th, 7th, and 8th chapters of Genesis, we have an account of the building of Noah's Ark, the Deluge, and destruction of mankind on account of their wickedness. That within the period of man's occupation of the earth a flood occurred, which swept off all the inhabitants, is an all but universal tradition. It is also a part of the same tradition that a family, with less or more attendants, were saved by means of a vessel prepared for the occasion. The recent discovery, by Mr George Smith, of a tablet in the British Museum, having an inscription upon it describing the Deluge, has given a new interest to the subject. This tablet has been found along with others in Assyria, and the various tablets probably formed part of the library of an ancient Assyrian king. The inscription is considered to record events derived from sources much older than the tablet itself. The tablet narrative agrees with the Biblical account in many most important points. The two statements affirm that the Flood was sent in consequence of the wickedness of the people; that Divine instructions were given for the preparation of a boat or vessel for the preservation of

the builder and his family, and also for the preservation of a number of animals for the restocking
of the world after the waters had subsided. There
are a few minor differences, caused, no doubt, by
the channels along which the tradition has flowed.
On this point Mr Smith says :— "The Bible
account is the version of an inland people. The
name of the Ark in Genesis means a chest or box,
and not a ship, and no notice of the sea or of
launching,—no pilots are spoken of,—no navigation is mentioned. The inscription on the tablet,
on the other hand, belongs to a maritime people.
The Ark is called the ship, the ship is launched
into the sea, trial is made of it, and it is given in
charge of a pilot."

Mention is also made of a great many people
being preserved in the Ark with the builder and
his family, a state of matters which, considering
the circumstances, we think necessary, and which,
although not contained in the Bible narrative, is
not excluded. It is not our purpose, nor is it our
province, to discuss the question whether the
Chaldean account of the Flood has been borrowed
from or founded upon the Mosaic, or whether the
account in Genesis has been from the same source
as that from which this Chaldean account has
been taken. According to Rawlinson, who is no
doubt a high authority on such questions, the date
of the tablet is prior to the time of Moses.

Our inquiry is concerned, not with the Deluge,

but with the Ark as a work of art. This is the
first great work of the construction of which we
have any record; and a careful consideration of
the details contained in the narrative is very suggestive concerning the state of practical and scientific knowledge to which the Antediluvian had
attained. With this end in view, we purpose
taking the Bible narrative, brief as it is, and
viewing the matter as much as possible from the
stand-point of a practical man. The object for
which the Ark was built is well known; and even
the most casual reader, with a very little reflection,
may see that God's announcement to Noah of a
coming flood, and His intention of saving him
and his family with pairs of all animals, entails
that one of two things was necessary in order
that Noah could properly execute his commission.
First, A full detail was necessary of the time the
Flood was to come upon the earth, the time the
water would continue, the number and kind of
animals to be preserved during that period, and
the circumstances in which the vessel was to be
placed during its voyage, to enable Noah to work
out the capacity and proportions of the required
vessel. For this purpose Noah would not only
require to be a good naval architect, but also to
be thoroughly acquainted with all existing animals,
their natures and habits, in order to provide them
with accommodation and proper food during a
five months' voyage and seven months' confine-

ment after the stranding of the vessel. *Second,*
That Noah, as the builder, should receive the size
and proportions of the vessel, with other necessary
details of internal arrangements; and when the
ship was finished, receive further instructions from
the divine Architect as to the amount of food, and
number and character of the animals, and other
details. This latter, apparently, was God's plan,
as the narrative leads us to believe that Noah
did not know the time when the Flood would
occur until after the Ark was completed; then he
was told to enter with his family and the animals
as specified :—" For yet seven days, and I will
cause it to rain upon the earth forty days and
forty nights." Even this could give Noah no
idea of the magnitude of the Flood. Noah, we
think, represents a period in which the art of
ship-building had attained a considerable degree of
perfection. He, as the master shipwright, wrought
to the plan laid before him, as is done at the
present day. There is no evidence that the workmen were inspired, or miraculously endowed with
practical skill in the use of tools. The instructions
Noah received for the building of the Ark, as
detailed by Moses, are as follows :—" Make thee
an Ark of gopher wood; rooms shalt thou make in
the Ark, and shalt pitch it within and without
with pitch. And this is the fashion which thou
shalt make it of: The length of the ark shall be
three hundred cubits, the breadth of it fifty cubits,

and the height of it thirty cubits. A window
shalt thou make to the Ark, and in a cubit shalt
thou finish it above ; and the door of the Ark shalt
thou set in the side thereof; with lower, second,
and third stories shalt thou make it."

Some of these directions are not very clear to
us, possibly from the translation not conveying the
real meaning of the original. Concerning the word
translated "Ark," the Rev. J. J. S. Perowne in
Smith's " Dictionary of the Bible," says, " The precise meaning of the Hebrew word is uncertain, the
word only occurs here, and in the second chapter of
Exodus, where it is used of the little papyrus boat, in
which the mother of Moses intrusted her child to the
Nile." Some scripture expositors read, " Windows
shalt thou make in the Ark," and consider that
the direction, " and in a cubit shalt thou make it
above," refers to the roof, or what we would call
upper deck, which was to slope a cubit. Others
again consider that the window was to be a cubit
in depth, running round the whole Ark at the
top, and formed a sort of skylight to the whole
interior, with its three tiers of nests or rooms, and
thus provided light and ventilation to the inmates.
Some again object to this idea on the ground that
to prevent rain getting in, glass or some other
transparent substance would be required. This
they consider could not be known at that early
period, forgetting that the skill required to con-
struct the Ark implies a state of practical know-

ledge, far beyond that which we look for in the
knowledge of a substance which will admit light
and keep out rain. Whether either of these be the
true view or not, there can be no doubt that proper
means were provided for ventilating and lighting
the Ark; and we cannot think that Noah and his
companions were shut in, as it were, under hatches.
They must have had means of taking observations
both as to the state of weather and water, as we
find the very day noted on which the tops of the
mountains became visible, and this before the
opening of the so-called window in the roof. And
it is not for some time after that Noah removed
the covering of the Ark, though what this covering
was is not stated, only it shows that all the details
of the vessel are not given in Scripture. As to the
general construction of the Ark, commentators have
followed each other from the days of Sir Walter
Raleigh with very little variations. Their views are
summarised in the following extract from Smith's
"Dictionary of the Bible," by the Rev. Mr Perowne:
" It should be remembered that the huge structure was only intended to float on the water,
and was not, in the proper sense of the word, a
ship. It had neither mast, sail, nor rudder; it
was, in fact, nothing but an enormous floating
house, or oblong box rather. As it was very likely,
says Sir Walter Raleigh, that the Ark had a flat
bottom, and not raised in form of a ship, with a
sharpness forward to cut the waves for better

speed, the figure which is commonly given to it by painters there can be no doubt is wrong. Two objects only are aimed at in its construction, the one was that it should have ample stowage, and the other, that it should be able to keep steady upon the water. It was never intended to be carried to any great distance from the place where it was originally built." We need hardly say that there are here a number of assertions and glosses not warranted by the narrative. The word translated " Ark " seems to be like the word vessel, which may apply to a box as well as a boat, that is, implies no particular shape. An order given to a boat-builder for a vessel to be 300×50×30 feet, one would never suppose that this meant an oblong square box, flat-bottomed, and without bow or rudder. Such a structure would not only be more unsuitable, but much more difficult to make water-tight than the usual form given to ships. The statement that " it was never intended to be carried to any great distance from the place where it was originally built," is entirely an assertion of the author, and altogether without foundation, as the Ark may have sailed a long distance during the five months it was afloat. Rawlinson, in his " Ancient Monarchies," speaking of the Ark, says— " Would a floating house, not shaped shipwise, have been safe amidst the winds and currents of so terrible a crisis ? " The author from which we quote proceeds, " A curious proof of the

suitability of the Ark for the purpose for which it
was intended, was given by a Dutch merchant,
Peter Jansen, the Menonite, who, in the year 1604,
had a ship built of the same proportions. It was
120 feet long, 20 broad, and 12 deep. This
vessel, unsuitable as it was for quick voyages, was
found remarkably well adapted for freightage. It
was calculated that it would hold a third more
lading than other vessels, without requiring more
hands to work it. A similar experiment was also
said to have been made in Denmark." It is not
said whether the Dutch merchant's ship was flat-
bottomed and without bow, or rudder, or mast,
only the proportions—rather a curious way of supporting the validity of an ecclesiastical gloss. We
have no evidence as to the truth of these opinions,
but rather consider them gratuitous, and such as
on a little consideration will be seen to be erroneous.
Nothing is said in Scripture about the shape of the
Ark, only its dimensions are given. However,
there is one remark in the extract given with
which we heartily agree—viz., the twofold object
evidently held in view in the design of the Ark,
ample stowage and steady sailing. In a large
cattle ship these qualities are most essential. We
quote a sentence from Dr Eadie's "Biblical
Encyclopædia," as an instance of unmeaning
description : "After the most accurate computa-
tion by those versed in ship-building, and supposing
the dimensions given in the Sacred History to be

geometrically exact, it is found that the vessel, in all its known parts and proportions, is in perfect accordance with the received principles of naval architecture; " and it is added, " there can be no doubt therefore that the Ark was built on strictly scientific principles." We think no believer in the inspiration of the Bible can gainsay this last remark, knowing who was the Architect.

We will now consider the size of the vessel. This is a subject of considerable difficulty, on account of the uncertainty concerning the exact length of the cubit which is the unit of measurement in the text. A cubit was the unit of length used by various nations, but the cubit of one nation was not of the same length as that of another. Some able scholars are of opinion that there are distinct cubits, varying from 18 to 21 inches in length, referred to in Scripture, this being the general length of a man's arm from the elbow. Whether, however, the arm should be measured to the wrist, or to the knuckles with the hand shut, or to the point of the fore-finger with the hand open, does not seem clear; hence there is a difficulty. There was also what is called the sacred cubit, which most of our commentators agree is that which formed the measure of the Ark, and of other structures which were made by Divine appointment. There have been several surmises as to what the length of this sacred cubit was. Sir Isaac Newton, as stated by Pro-

fessor C. P. Smith in his "Work at the Great
Pyramid," after a careful investigation, calculated
its length to be 25·025 inches; his reasons for
adopting this measure being that it was the ten
millionth part of the earth's semi-polar (or semi-
axis of rotation) diameter. We are strongly
inclined to favour the views of Sir Isaac in this
matter. However, let us take each of the three
different kinds as unit, viz., 18 inches, 21 inches,
and 25·025 inches, and we have the following in
feet—

18 Inches Cubit.	Length of vessel,	.	.	450	feet
„ „	Breadth „	.	.	75	„
„ „	Depth „	.	.	45	„
21 Inches Cubit.	Length of vessel,	.	.	525	feet
„ „	Breadth „	.	.	87½	„
„ „	Depth „	.	.	62½	„
Sacred Cubit.	Length of vessel,	.	.	625½	feet
„ „	Breadth „	.	.	104¼	„
„ „	Depth „	.	.	62½	„

This last measurement is only 55 feet shorter
than the *Great Eastern,* her length being 680
feet; even the lowest is a vessel equal to the
average of any we build at present for long sea
voyages.

As no art or manufacture begins in a high state
of perfection, but is a gradual development, we
think Noah stands upon the apex of a long
growth in the art of ship-building; for although
God is represented as the Architect, Noah is the

person directing the work. If such an order as he received had been given to a people in the rude state of ship-building represented by the many ancient canoes found in our own country, it would not only have been unmeaning, but, without miraculous aid, impossible to carry out. No hint is given by the inspired writer of any difficulty having been experienced in the construction of the vessel.

The next important consideration is the proportions of the Ark, and from this we can form more definite conclusions. When a naval architect is asked to design a vessel, he must know previously the purposes to which the vessel is to be put; whether she is to sail on the ocean or on a river. A vessel made merely to float upon still water, with heavy cargo, will be quite differently proportioned from one carrying a similar cargo, but intended to battle with wind and waves. If the Ark had been merely intended to float in still water, she would have held far more cargo by an increase of breadth, but we think from her dimensions and proportions the Ark was made suitable for large cargo and heavy sea. If a naval architect of the present day were asked to design a vessel for the purpose of carrying large cargoes with the greatest speed and capable of encountering heavy seas, he would design one in the proportions of our Clipper ships, or something approaching these proportions. To show what these were and are,

we quote from a work on Naval Architecture by
the late Professor Rankine and J. R. Napier :—

"The common proportions of length to breadth,
in navies of the world down to recent times, ranging
from about 3 to 3¾ and 4, was considered a great
proportion even in yachts. At present many
sailing vessels are built longer and narrower than
these proportions, but still in the finest example
of Clipper ships, the length is seldom more than
5 to 6 times the breadth."

Here we find that both in old and in more
modern times, before naval architecture was so
well understood as it is now, vessels built for great
capacity and sailing power had nearly double the
proportions of breadth to length compared with
what modern science and practice say are the
most suitable proportions. Shall we then say that
the fact of the length of the Ark being exactly six
times its breadth, the proportions found to be most
suitable for quick sailing and a strong sea was
a mere accident. We are rather driven back upon
our statement, that the design of the Ark was inspired, and the work carried out by men who were
skilled in work of the kind. Suppose our friends
the archæologists were to find the remains of such
a vessel as the Ark, we think they would reason
from the stand-point of a naval architect, that the
people in the age in which it was built must have
been far advanced in the art of ship-building. Yet,
as we have seen, Biblical commentators generally

do not perceive the scientific perfection in the structure of the vessel, but make of it something as ridiculous as it is well possible to do from the data given. Looking at the whole structure from a workman's point of view, there would be required not less than 25,000 loads of timber for its construction. The cutting down of such a number of trees as would be required, their conveyance to the building-yard, the preparation of the wood for the timbers of whatever shape they were, and the fitting of the different parts, indicate a great amount of skill on the part of the workmen. Whether the fastenings were of wood or metal we of course cannot decide, but considering what we have seen of their knowledge of metals in ancient times, it is more than likely both were used. A recent correspondent of the *Athenæum* says that the knowledge of metals was retained by the descendants of Cain, and the reason why Noah was so long in the building of the Ark was that he would have only flint or other stone implements or tools to work with. This certainly would make the work more difficult, but it does not detract from the skill of the workmen; rather does it speak highly for their skill and patience in undertaking and finishing the work under such adverse circumstances. However, we are not inclined to favour such an idea, but think that metal tools were in the hands of the Ark builders; for, if the workmen employed by Noah did not know

metals before the flood, they were not long in
making the discovery after, for we find metals in
common use at dates immediately after that given
for the Flood among the immediate descendants of
Noah. Neither do we think that one hundred and
twenty years were spent in its construction. All
the Bible says is, that it was one hundred and twenty
years from the time Noah received his instructions,
till the Flood came. We have heard objectors say
that the wood used at the commencement would be
rotten before the vessel was finished, forgetting that
provision was made against such a result, by having
the wood saturated with tar, or in the language of
the Bible, "pitched within and without."

There are some commentators who are so inno-
cent of anything practical in matters of this sort,
as to consider the Ark to have been made of basket-
work made watertight by pitch, because the same
word is used in the original which is used in
reference to the little boat (probably of basket-
work) in which the infant Moses was placed in
the Nile. We have already shown that the Ark
was a regular ship, that it was necessary to be of
great strength to support the large cargo it was
to carry, and we will show that it required both
stearing gear and means of propulsion to enable
the voyagers to manage it. We are aware that in
stating this we are treading on very delicate
ground, and that the orthodox view is that God
preserved the vessel and cargo by means of a

z

continued miracle. But this is contrary to God's way of dealing. There is no instance in Scripture of God interfering by miracle where man, under the ordinary laws, can effect the object. He had instructed Noah in all that was required in the vessel, for the comfort and protection of the different animals, by having them placed in different rooms or compartments, and also for the proper working of the vessel, and Noah was no doubt left to manage all the rest himself. The Ark was afloat for five months, and it is hard to believe that such a catastrophe as the Flood was unaccompanied by wind; at all events, we cannot suppose that there were neither wind, waves, nor currents during the whole five months' voyage. A vessel without rudder in currents and wind always drifts side on. A vessel, such as the Ark, without rudder or means of propulsion, placed in a current accompanied even with slight wind, would be placed in a situation not only most uncomfortable, but one in which a miracle would be required to preserve alive a cargo of cattle; so that we are driven to one of two alternatives, either the preservation by miracle, or, as we have explained, a vessel fitted up in such a way as to enable Noah to manage her by ordinary means. But, say the objectors, How could Noah and his three sons and their wives—eight persons in all—manage such a vessel by the ordinary means? This same difficulty remains, although Noah had not to work the

ship, as the animals were both to feed and keep
clean, which, considering their number, was more
than any eight persons could accomplish. The
simplest explanation of the difficulty is that we
have already suggested, that Noah and his sons,
like Abraham and the other patriarchs, are named
as representing their households, and these included
a great many servants, both male and female.
Indeed, the invitation into the Ark suggests this :
" And the Lord said unto Noah, Come thou and
all thy house into the Ark ; for thee have I seen
righteous before me in this generation." There
are many instances of this use of language in the
Old Testament history, where only the head of the
house is named singly, but historically he represents his whole family and adherents, and the
same phraseology is used in the New Testament
in the same wide sense. The Chaldean tablet to
which we have referred states that there were
many people along with the person named—the
Noah of the Bible ; and as we have shown that this
is not only not contrary to the Scripture account,
but essential to it, we accept it as strong evidence
of the correctness of our opinion.

The description given of the approach of the
Flood forces on our mind the necessity of Noah
having had control over his vessel. We are aware
that poets and preachers have represented the waters
gradually but steadily rising, and the people in
terror ascending the mountains in vain hope of

evading the rising tides, while they beheld the
Ark calmly floating away in safety with its living
freight;- and when now too late to escape their awful
doom, they in despair curse their wretched fate
and vainly regret their want of faith in Noah's
warnings.   Such ideas are no doubt full of poetry,
and touch the feelings of the hearers or readers.
Nevertheless, we think the beauty of the picture
is obtained at the expense of truthfulness.   The
great master Teacher referring to the Flood, com-
pares it to the coming of the Son of Man, which is
always represented as being sudden and unexpected.
People were marrying and giving in marriage
until the very day Noah entered the Ark.   And the
narrative as given by Moses is terribly grand, and
gives no sanction to a slow and gradual rising of
the waters.

"In the six hundredth year of Noah's life, in
the second month, the seventeenth day of the
month, the same day were all the fountains of the
great deep broken up, and the windows of heaven
were opened.   And the Flood was forty days upon
the earth, and the waters increased and bare up the
Ark, and it was lifted up above the earth.   And the
waters prevailed and were increased greatly upon the
earth.   And the Ark went upon the face of the waters,
and the waters prevailed exceedingly upon the
earth."   In the marginal reading of the A.V. the
words, "The waters prevailed," bare the meaning,
"Bearing down everything before it;" and such

would certainly be the case. Mr Steinmitz, in his book on "Meteorology," says, "It would be a tremendous shower if the rain fell at the rate of two-hundredths of an inch per minute for one hour;" and he makes a calculation based upon the statement that the water rose to the height of fifteen cubits above the highest mountain in forty days and nights, and finds that the water would require to rise at the rate of five inches per minute, or to twenty-five feet per hour. This would be sufficient to destroy every living thing in the space of half an hour. Without adopting the calculation of Steinmitz of a rise of five inches per minute, but only supposing the rain to fall at the rate of two inches in an hour, it would be impossible for any one to climb a hill. Such a shower would sweep down everything movable before it, burying the valley in one heap of confusion; so that the idea of poet and preacher is contrary to fact, and is tame when compared with the historical record.

However, it may be said that if ship-building were common, it is more than probable that many other people besides Noah were in possession of vessels and could have saved themselves. Many would be afloat in vessels following their trade when the Flood came, and might be saved, so that in order to meet the whole case it is necessary that the Ark should be the only and probably also the first ship ever built, as commentators will have us believe. We admit that if ship-building was a common art, there would be many more

vessels in possession of men, who could save
themselves, and perhaps did save themselves from
the first great downpour of the Flood; but if we
take the "breaking up of the fountains of the deep"
to mean a rapid raising of the water as an immense
wave, or a succession of waves, how many vessels
could survive such? and if any did survive few would
be provisioned for twelve months.   Mr George
Smith, speaking about the Chaldean tablet, says
that the Scripture account evidently refers to a
vessel built inland; there is no account of launching
it, &c.   We agree with this supposition, and think
it is Scriptural.   The Ark was built for a special
purpose and under special circumstances, and
did not require to be launched to test its fitness;
that was insured by the Architect.   It is more
than probable that the vessel was built upon some
inland plain, contiguous to a plentiful supply of
wood.   And the fact of building such a vessel
away from the sea may have brought upon Noah
the derision of his neighbours, as also that the
people refused to listen to his warning, which has
been traditionally connected with the undertaking.
Indeed, who would put their trust in the preaching
of a man whose conduct was apparently so absurd?
When finished, this inland position was the best
for collecting the cargo to be saved in the ark.   It
shows the necessity of the door being in the
side, which could be afterwards secured, and the
securing of this may be what is referred to by the

expression, " God shut him in." And again if the
great catastrophe referred to, " the breaking up of
the great deep," was, as we think, sudden, whether
by earthquake or by the sinking of the land, few
if any vessel afloat could survive the fearful reflux
and influx of waves which must ensue. Even
within our own day, in South America, earthquakes
near the sea have been destructive to everything
near them, and all vessels within the influence of
the great wave have been destroyed, and these
were but slight compared with the Flood. Consequently the building of the Ark inland secured its
safety, being protected from all destructive effects.
In the language of the Bible, it was borne up on
the surface of the waters.

. There is one other statement in the narrative of
Moses which has led to some difficulty: that is the
expression we have referred to above, viz., "That
God shut him in." We have put the question to
several clergymen, Hebrew scholars, whether
these words mean that God closed up the Ark,
and that the inmates sat within passive so far as
the vessel was concerned. Four out of six held
this opinion, and that all were preserved by
miracle; two said the words so translated did not
necessarily bear that meaning—that so far as that
expression was concerned, Noah might have
wrought his vessel as any captain would his
ship. But there being no apparent destination,
the guidance would only be to keep it in the

most favourable position for the comfort of the inmates.

One German author whose book was recently reviewed in the *Athenæum*, cut the Gordian knot at once, by saying that God put all the inmates of the Ark into an hibernating state, and awakened them after the waters were abated; but to ascertain if the waters were certainly abated, Noah sent out the dove and raven, and after being convinced by the dove not returning, preparations were made for removal. We must confess that if miracle is admitted in this matter at all, this view is the best. But unfortunately for such ideas, the narrative viewed in a practical manner gives us no warrant either to seek for or believe in a miraculous preservation of the Ark. After providing Noah with a suitable vessel, that God should afterwards be necessitated to work a miracle on account of insufficient manual labour or skill on the part of Noah, seems inconsistent with His perfect character.

There are several other points in the narrative, which we think require a different explanation from that generally given, to be consistent with the authorship, and with common sense. But these are outside of our subject, which is with the Ark itself, and its construction. We will conclude this chapter with a quotation from Baldwin's "Prehistoric Nations:" "The nautical science of the old Arabians must have been equal to the

wants of their great commercial enterprise, for the
faculty to invent what was necessary did not wait
until our time to make its appearance in the
human mind. It is seen that the Phœnicians,
when they came out of the obscure prehistoric
ages into history, were immeasurably superior to
every other people in maritime skill. No other
people had such naval constructions, no other
people were so much at sea, or made such long
voyages, or had such skill in navigation. Some
ages later, the Athenians had ships, but they had
nothing like the great Phœnician ships that came
regularly to Athens in the service of commerce,
and were always regarded there as marvels of
naval architecture. What the Phœnicians were in
the Mediterranean, the southern Arabians were on
the Indian Ocean, although history necessarily has
much less to say of them. And these people, representing in the later ages of its decline the great
race that had been for not less than six millen-
niums, probably, foremost in human affairs, give us
some notion of what must have been the attainments of that race in the ages of its highest
development and power. The marvellous skill of
the Phœnicians in manufactures would have been
impossible without unusual intellectual activity,
aided by great attainments in science. Is it
reasonable to believe that the nautical science of
such a people, with whom maritime enterprise was
a chief business, could have been poor and mean?

Such a belief is possible only to the prodigious incredulity of egotistical scepticism."

The same author is also of the opinion that the ancient Arabians knew the mariner's compass. He says, "The word *vorsoriam* has been interpreted to mean the mariner's compass, although some critics have tried to see in it a rope, or the helm of a ship. Pineda, and Father Kircher, argued earnestly to show that the compass was used by the Phœnicians and Hebrews in the time of Solomon, which was much more reasonable than the claim that it was first invented in Europe."

The Chinese, too, are said to have known the magnetic needle more than 2700 years before the Christian era.

INDEX.

THE END.

PRINTED BY BALLANTYNE AND COMPANY
EDINBURGH AND LONDON